Volcano Cowboys

DICK THOMPSON

Thomas Dunne Books

St. Martin's Press   New York

Volcano Cowboys

The Rocky Evolution of a Dangerous Science

Thomas Dunne Books.
An imprint of St. Martin's Press

www.stmartins.com

Book design by Kate Nichols

The photographs in this book were prepared as part of official duties of the U.S. Geological Survey; therefore, the photographs are in the public domain and no permission is required for their reproduction or use.

Library of Congress Cataloging-in-Publication Data

Thompson, Dick.
   Volcano cowboys : the rocky evolution of a dangerous science / Dick Thompson.
     p. cm.
   Includes index.
   ISBN 0-312-20881-2
   1. Volcanological research—History—20th century. I. Title.

QE521 .T48 2000
551.21'072—dc21   00-026158

First Edition: July 2000

10 9 8 7 6 5 4 3 2 1

For Dave and Harry
and for K

Contents

PART III: MOUNT PINATUBO, 1991

Volcano Cowboys

Introduction

Volcanoes are magnificent primordial beasts. They are geology's living dinosaurs. They reveal how the earth has shaped and recycled itself for millions of years. And they have molded beautiful landscapes—breathtaking mountains, snowcapped peaks, plunging rivers, tropical islands, and glistening lakes. Volcanic lands are often so beautiful and their ground so fertile that they entice people to their shaky shoulders. This, plus the encroachment of suburbanization, means that tens of millions place themselves within the reach of volcanoes every day.

Occasionally, these picture-postcard places become landscapes of almost unimaginable violence. Volcanoes have ripped up whole forests, down to the bedrock, and scattered the trees as if they were autumn leaves. Volcanoes have swept aside lakes, blown mountaintops miles into the sky, launched shock waves around the world, spawned tidal waves that killed tens of thousands, and altered the earth's climate.

Any good bookstore has a hefty shelf of books describing the science of volcanology. This isn't one of them. This is a book about how science operates. It is the story of how one scientific goal, the prediction of explosive eruptions, evolved during its most intense and productive periods, from the eruptions of Mount St. Helens to the cataclysm of Mount Pinatubo. I have drawn from those other science books, and I have been lucky enough to have had many of the authors share their expertise, experiences, sources, and occasionally even their field notes with me.

This is a story of how science operates in periods of crisis. The pressures placed on scientists when volcanoes threaten can be crushing. Academic collegiality, presumptive prestige, and the myth of scientific power are most likely to be stretched thin. The raw process of science is exposed. The high-tech gizmos are scrutinized, and technical language is translated. Economic and safety concerns have to be addressed. Careers are on the line. Nonscientists from loggers to generals discover how much, or how little, faith they really have in the priests of reason. Many volcanologists I interviewed say a volcanic crisis is a pressure-cooker environment. It's something like the state of a volcano on the verge of exploding, a condition volcanologists refer to as overpressure.

Like other scientists, volcanologists have their labs—the volcanoes themselves. The great Australian geologist G. A. M. Taylor called volcanology the Cinderella science because it "only marches forward on the ashes of catastrophe." So reawakening volcanoes are a scientific feast for volcanologists.

These great test tubes pose unique problems. First, volcanoes don't perform on cue. For most of volcanology's history, few scientists actually witnessed the initial eruption of an exploding volcano. They had to wait for word that one had blown, then they would hop a ship—usually to some exotic land—arriving weeks after the event to record their observations of the destruction. Even today, while the science is far better at identifying volcanoes that could soon explode, there is never a guarantee they will follow the script. Scientists can attend a pregnant volcano for weeks or months as it pressurizes, only to have it, for some unknown reason, deflate without a puff.

Because the object of study is so unpredictable, volcanology has advanced in spurts, with great droughts linking brief outbursts of productivity. This pacing left this field, perhaps more than any other, dependent upon the literature.

Dependent upon the literature—that's a phrase often heard in science. Scientific literature represents the work and often the hardship of the scientists who went before, some of whom risked their necks and sometimes their families as well. Volcanologists are deeply rooted in the history of their science to a degree unlike those in most other fields. The vital memory of molecular biologists, for example, generally drops off after a decade, two at most. Volcanologists will beguile you with stories about the roles of Greeks and Romans in their studies.

And the ancients did play important roles in volcanology. Plato was the first to describe the source of lava. Aristotle named the depression in the tops of volcanoes a *crater* for "cup." The Greek philosopher Empedocles may have been the first victim of the science. He supposedly sat down on the slopes of Mount Etna to absorb its mysteries, but when that didn't work, legend has it, he threw himself into the fire.

Others made more substantial contributions. Seneca, who tutored the Roman emperor Nero, was the first to describe the role of gases in an eruption. Seneca was the first to suggest that each volcano is fed by its own reservoir. He also theorized that underground fires excite gases that tighten "the springs of the blast" until they are released by the explosion. It was a stunning insight that took about two thousand years to confirm. And in his astounding thirty-seven-volume encyclopedia, the Roman naval officer Pliny the Elder included a list of the known volcanoes in the world; at the time they numbered ten. One of them would kill him.

His nephew Pliny the Younger would carefully record the death of his uncle and many other details of the A.D. 79 eruption of Mount Vesuvius. His two remarkable letters are truly the first pages of descriptive volcanology. He detailed the foreshadowing earthquakes, the enormous vertical column of ash that stands like an Italian parasol pine, the thick lightning that accompanies ash clouds, the sulfurous gases, the pumice that buries buildings and fills the lungs, the hot ash flows, and the receding of the sea, leaving fish flopping on the exposed seashore, that preceded the tidal wave.

Pliny the Younger's uncle may have been asphyxiated by the gases spewing from the volcano or killed by the thick clouds of ash, either would have been as hot as the inside of a kiln. With his uncle dead and their home shaking apart, Pliny and his mother joined the crowds of refugees fleeing Vesuvius. On the road, they were plunged into darkness, although it was the middle of the day.

"I looked behind me," he wrote of his escape. "Gross darkness pressed upon our rear and came rolling over the land after us like a torrent. I proposed while we yet could see, to turn aside lest we should be knocked down in the road and trampled to death in the dark by the crowd that followed us. We had scarce sat down when darkness overspread us, not like that of a moonless or cloudy night, but of a room when it is shut up, and the lamp is put out. You could hear the shrieks of women, the crying of children, and the shouts of men. . . ."

Pliny was what volcanologists continue to admire most, a threatened observer who recorded with precision the details of an eruption. Geologists expanded on Pliny's descriptions, but they never rewrote them. In tribute, the towering vertical column of ash and gas exploded in catastrophic eruptions is called a Plinean column.

The next important volume in the volcanic literature comes eighteen hundred years later. It detailed the 1883 eruption of Krakatau, which generated the loudest sound ever heard and a tidal wave that accounted for most of the thirty-six thousand people the eruption killed. It was also the first time modern communication, the telegraph, helped to identify the global consequences of a volcanic eruption, primarily the brilliant red and orange sunsets that around the world were confused with nearby conflagrations.

By the end of the 1800s, scientists began concentrating on the components of an eruption. For example, pyroclastic flows, the fast-moving, boiling clouds of ash, were first described by volcanologists in 1873 at Santoríni, Greece.[1] Pyroclastic flow studies began in their modern form with Mount Pelée's eruption in 1902 on the Caribbean island of Martinique. In that eruption, the rapid flows swept along the ground, ripping through the beautiful tropical city of Saint-Pierre and, within five minutes, killing all but two of the city's twenty-nine thousand residents. But scientists had some good fortune. Pelée emitted subsequent pyroclastic flows that provided opportunities for them to study the phenomenon.

One of the best-known scientists to study the pyroclastic flows on Martinique was an inventor of electrical motors and a former assistant to Thomas Edison named Frank Perret. Perret's courage and ingenuity are legendary. At Pelée, he monitored earthquakes by biting a brass bedpost that had been set in cement. He found that his teeth detected even the slightest tremor.

For his studies of pyroclastic flows, Perret worked for two years on the side of Pelée, sometimes from a hut less than a hundred feet from the flows. His hut was once overwhelmed and Perret himself was enveloped on the fringe of a pyroclastic flow. He is one of the few people ever to have survived such an encounter. He was also the first to report the sound pyroclastic flows make as they move at hurricane speed: they make no sound at all.

After the eruption of Mount Pelée, it became obvious to American scientists that if volcanologists were to break free of the constraints of serendipity, they needed a volcano that erupted frequently. U.S. scientists found one, actually two, in Hawaii: Mauna Loa and Kilauea, both on the Big Island of Hawaii. The U.S. Geological Survey's Hawaiian Volcano Observatory is still the primary training and testing ground for the Survey. But while several generations of American scientists trained on those volcanoes right up to the time of the 1980 eruption of St. Helens, the Hawaiian volcanoes had one deficiency. They were not as explosive as the great stratovolcanoes such as St. Helens.[2] Compared to Krakatau or Pelée, in modern times the Hawaiian volcanoes have only erupted quietly with slow-moving rivers of lava oozing down their flanks. This turned out to pose a particular problem for the Survey scientists who rushed to Mount St. Helens in the late winter of 1980. St. Helens was a different animal.

One last feature of volcanology has made it uniquely difficult for the field to advance—the volcano lab can be a bit like working in a minefield. Doing this science can cost a researcher his or her life. In 1993, an eruption occurred while volcanologists were working inside the crater of Colombia's Galeras. Six of the scientists and three others were killed.

While most of us would run a safe distance from an exploding volcano, volcanologists are generally rushing the other way. The only restraint on their curiosity has been their well-reasoned fear, although they generally don't use that word. One volcano scientist told me it was an attempt at a "realistic appraisal of risk versus benefit." But, in fact, in most of these situations, both the risks and benefits are largely unknown. So there is no hard line in any volcanic situation. It's always a personal calculation, and the factors in that calculation can change in a minute. Often, a scientist's curiosity has trumped fear, and sometimes the consequences have been tragic.

This story focuses on the scientists of the U.S. Geological Survey. There are many outstanding volcano experts elsewhere, both in the United States and abroad, and I hope singling out the USGS volcanologists will not appear to minimize the contributions of others. Their work will stand as bricks in the edifice of volcanology long after this book is recycled and recycled again.

I chose this focus because, beginning with St. Helens and continuing to the eruption of Mount Pinatubo, only Survey scientists were able to devote the months of concentrated work in the field that these events required. Moreover, once Survey specialists responded to a crisis in the United States, they sometimes deliberately limited participation by any outsiders. They often saw other scientists as distracting them from their main responsibilities—to monitor, to learn, to predict, and to warn of a threatening volcano. The other scientists, of course, believed the Survey personnel just didn't want the competition.

This closed system worked well for the Survey in another way. It helped maintain the Survey's credibility. While the government scientists had differences of opinion among themselves, these differences rarely spilled outside the Survey. Thus, the Survey generally spoke with one voice. Dissenters were outsiders. It is a paternalistic approach to public policy making and fundamentally undemocratic. But it is an approach common to many sciences.

I have focused on the period between the eruption of Mount St. Helens in 1980 and the eruption of Mount Pinatubo in 1991. Compared to the rest of its history, the science of volcanology moved at warp speed in this time. Beginning with the 1980 eruption, American volcanologists had an explosive fire-breathing monster to observe and measure and poke. Following the catastrophic eruption in 1980, Mount St. Helens had six more explosive eruptions and many more non-explosive, dome-building eruptions. The beast didn't return to sleep until 1986.

In addition to benefiting from the pace of eruptions, Survey volcano experts in the 1980s also became better funded than they had ever been. Before 1980, they had to fight for survival budgets, but for a brief period afterward they had all the resources they needed (although never as much as they wanted). After the May 18 eruption, they also had the incentive and the encouragement to go abroad and study explosive volcanoes outside the United States. All of this opened a new era in the science.

I have been able to talk to nearly all of those involved in this period. Often, I have been able to talk to them for hours. These interviews were supplemented with newspaper clips, project proposals, telephone logs, memos, published scientific articles, drafts of memos, and sometimes even the scientist's field notes. Rocky Crandell gave me the two volumes of his unpublished autobiography. And nothing matched Rick Hoblitt's Pinatubo field notes. They are a work Pliny the Younger would admire.

I want to thank them all for giving me so much of their time. Just meeting some of them (such as Ray Wilcox, who was the first permanent observer at Parícutin in 1948) became a reward in itself for doing this project.

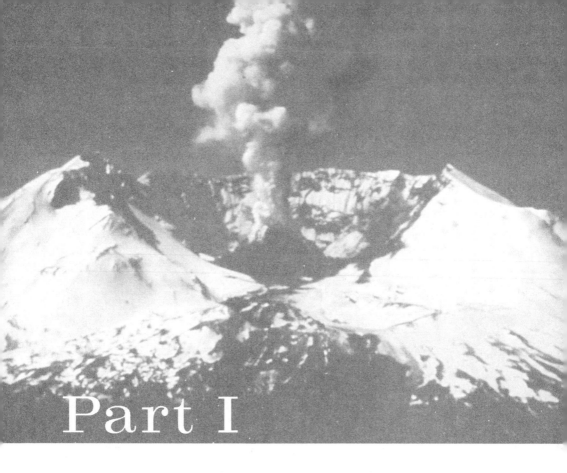

Part I

Mount St. Helens, 1980

CHAPTER 1

Hoblitt's Floating Island
Summer 1979

Just below the timberline, where the snowfields stopped and the forests began, at an altitude of forty-five hundred feet on the north flank of Mount St. Helens, Rick Hoblitt laid down his cantankerous rock drill and his overstuffed backpack. He was at the edge of a forest, near a rock field the size of Central Park. Hoblitt, a scientist with the U.S. Geological Survey, and his summer field assistant had spent most of this entire August day in 1979 hiking through the forest and hacking through the brush to reach this odd gash of rubble called the Floating Island Lava Flow. The two men built their camp beside a fast, cold stream that ran milky white with rock flour from one of the summit's glaciers. And as the chill wind began to sweep down from the snow-fields, they built up the campfire until the skillet was hot enough to sizzle when Hoblitt slapped down two slabs of meat in it.

Pilots flying fishermen into the lodge at Spirit Lake could easily spot the boot-shaped slab of the mountain. But you had to be as close as Hoblitt was now to understand why almost forty years earlier a botanist had named it Floating Island. Close up, it was a great, choppy river of lava frozen in sharp, black boulders. Scattered here and there in the rock river were islands of soft earth where small groves of trees stood. It looked as if, far uphill, pieces of forest had broken away and these green islands had rafted down the black lava river.

Like the rest of St. Helens, the Floating Island flow was young, by geological calendars. Just beneath these boulders was a solid, massive rock that had oozed from its vent sometime around 1800. Hoblitt had hiked into this area several times to drill samples of rock from the inner core and to take plugs from trees. Both samples would help pin down the precise date of the particular eruption that created this feature of St. Helens.

Hoblitt also planned to hike farther up the mountain, along the edge of the Floating Island flow to Goat Rocks and its fan of rubble, to drill more samples. Goat Rocks was a lava dome, and to a volcano specialist such as Hoblitt, it was obviously another spigot of the volcano's plumbing system. A century ago, magma had risen through some ancient passageway and flowed out the vent until the weight of the cooling surface lava became greater than the upward pressure of the rising magma. At that point, the pipeline was corked by a mushroom-shaped mound of fresh, plastic lava. The stuff hardened into rock on the surface. Geologists called these formations plug domes. St. Helens had five others in addition to Goat Rocks. Each one was a billboard pointing to a magma vent. Inside the mountain, each vent led back like the limbs of a tree to a central conduit that dropped more than five miles below the surface into a reservoir of liquid rock.

Hoblitt loved this region of the volcano. Here, next to the islands, the mountain was not only physically beautiful, it was littered with mammoth reminders of St. Helens's young, hyperactive life. In the volcano's recent years, mud floods called lahars had repeatedly rushed through the forests near Hoblitt's camp. These floods were created when St. Helens erupted lava of eighteen hundred degrees Fahrenheit onto the snowpack and glaciers. As the flood sped downhill, it mixed with topsoil and absorbed boulders, then churned through forests, adding trees to the mix. The tumbling mass became as thick as wet concrete. There was evidence of lahars on trees near Hoblitt's camp. Their bark showed old scars on the uphill side. But the most convincing piece of evidence was a boulder the size of a Volkswagen suspended twenty feet off the ground between two tall trees.

At the lower end of the Floating Island flow, three miles from the vent, there were signs that the lava flowing in the early 1800s had begun to cool as it came to rest. Here, majestic trees, mostly Douglas fir and

Western red cedar, had been pushed by the lava into various unnatural positions, like logged timber stopped midway in their fall.

As lovely as the view was, working here was a challenge. The rocks on the surface of the Floating Island flow were so young that they hadn't had time to settle. So footing was unstable. If you weren't careful, your boot could get wedged between the shifting boulders. Another thing that made the work difficult was the cheap rock drill Hoblitt was forced to use. Most of the time, it was in Harry's Small Engine Repair shop in Vancouver, an hour's drive southwest. And when it did work, he had to carry gas and water for the drill into the woods. But when things got too hot or unpleasant, Hoblitt would pick his way across the rocks to one of the cool, shaded islands and take a nap.

You didn't need to be a geologist to appreciate St. Helens. From many viewpoints, the mountain was a wildlife pinup, steep, with a sharp, snowcapped peak. The upper cone was always swaddled in snow and silvery glaciers, and the lower flanks rippled with thick green forests. In fact, St. Helens was so beautiful that it was regularly used by professional photographers for postcards and calendars.

The most romantic picture showed a fisherman standing in his boat, rod bent toward the water, with the peak of the mountain reflected almost perfectly on the still surface of Spirit Lake. The lake itself was clear and cold and filled with steelhead and salmon. The forests that ringed the lake were so thick that in many places sunlight could not fall directly on the forest floor. It scattered in the treetops creating a diffused, gentle light. Bands of deer and elk, along with an occasional coyote and black bear, moved slowly in the dappled light. The spongy trails were lined with ferns, and every breath of air carried a pine scent.

When the Lewis and Clark expedition came within sight of the mountain in 1805, William Clark described it in his leather-bound journal as "the most noble looking object of its kind in nature." St. Helens was supposedly the birthplace of Big Foot and was the suspected resting place of the first airline hijacker, D. B. Cooper. But for climbers, the mountain had a nasty reputation. It was quick to avalanche, and the loose material beneath the snow made footing unsure. Still, they liked the mountain. The 9,677-foot peak could be a good one-day training climb for average mountaineers.

From the air, one could see that St. Helens was not an untouched nature reserve. Much of the area around the mountain—in every direction but north—had clear-cuts made by the Weyerhaeuser and Burlington-Northern logging crews. The result was a vast checkerboard pattern that reflected the complicated land holdings in the region. But a wide swath to the north had been left nearly pristine from the mountaintop to Spirit Lake and beyond. Most of the loggers had spent their childhood summers on Spirit Lake, and they would not have tolerated even Mr. Weyerhaeuser stripping their memories.

The mountain's reputation as a volcano was enhanced by local storytellers, such as those who owned the Spirit Lake Lodge. A copy of a painting depicting the 1857 St. Helens eruption was strategically hung in the lodge's dining room as bait for newcomers. It was a strange-looking eruption, with a sparkling fountain of lava blowing vertically about midway up St. Helens, at the Goat Rocks dome. Mark Smith, whose family owned the lodge, would spin tales about how dangerous the mountain was.

"We'd give the whole spiel about it," says Smith. "But it was strictly dinner talk. None of us expected anything would ever happen. We all considered that mountain dead."

As Hoblitt watched the meat sizzle and smoke, he was focused on the volcano's past, which was his job and his joy. St. Helens's history was scattered all about Hoblitt's camp, and it was the work of his Survey team to piece it back together.

Time and again, eruptions from its summit had sprayed hot rock that melted the snowfields and launched great floods of muck fifty feet high. At other times, the volcano had burst with scorching winds of ash and pumice known as pyroclastic flows. The only living things likely to survive a pyroclastic flow would have been the pocket gophers hiding in their burrows. Spirit Lake, which stretched out at the volcano's northern foot, was itself the product of an eruption.

Five hundred years earlier, an eruption filled a narrow valley at the base of the mountain. The debris dammed a stream that pooled over the next one hundred years into Spirit Lake.

St. Helens had erupted so many times, in so many ways, that geologists had long ago stopped thinking of it as a solid mountain like the great slabs of the Himalayas and the Rockies. Rather, St. Helens was formed as the result of a series of eruptions. Technically, mountains like this are called composite volcanoes. Magma sheets spread on top of plug domes on top of rock rubble, and so on building a mountain whose internal structure is something like a pile of clothes.

During these summer field seasons, Hoblitt spent much of his time studying the deposits from old eruptions and drilling rock samples from hundreds of spots in the volcano. He carefully mapped the location of each rock plug, taking special care to note the orientation of the sample. He took those samples back to his group's lab in Denver where, over the fall and winter, the samples revealed details of the volcano's eruptions. He could tell how hot a glob of lava was when it was emplaced. He could even distinguish between rocks that had tumbled into place, cold from a lahar, and those that had arrived hot from a pyroclastic flow. Each sample was a piece of the puzzle, actually a three-dimensional puzzle that documented the volcano's history. Now after five years of putting the pieces together, Hoblitt could almost see the blasts and feel the heat. And that may have been the most frustrating part of his job. He wanted to see his volcano erupt. But given the whims of any volcano, he might spend his entire career working on this mountain and never see so much as a puff of steam.

Otherwise, life for Rick Hoblitt in the summer of 1979 was almost perfect. He had not suffered from his strange academic path, as he once feared. And he was working for the U.S. Geological Survey, the world's premier geology outfit. That last quiet summer field season, before his children were born and before St. Helens became a headline and a tombstone, Rick and his wife, Marian, lived in a leaky, white, twenty-two-foot house trailer parked in the woods, near the shore of Spirit Lake. They had no television and the nearest phone was a twenty-minute drive away. Inside the trailer, their cat curled up in the sunlight. When it rained, and it rained often that summer, Rick and Marian walked through the forest identifying mushrooms. At night, they visited with their neighbors, most of whom worked with the Forest Service. And occasionally the couple hiked two miles for dinner at Harmony Falls.

Hoblitt's odd career path had led him here. In the early seventies, while picking up his master's in chemistry, he had stunned his academic advisers at the University of Colorado by abandoning chemistry for geology. He told Marian he had to do it. The idea of spending his life in a chemistry lab, walled off from the outdoors in a room stinking with fumes, seemed too bleak.

Just as his dissatisfaction with chemistry had reached a boiling point, Hoblitt had been seduced by the prospect of becoming a field geologist. The idea was planted by Dan Miller, a neighbor of the Hoblitts' in Boulder's married-student housing. They had become friends almost from the day Miller arrived on campus in his geriatric Cadillac convertible. The two students were very different. Hoblitt was a quiet, compact intellectual with long brown hair and a dry sense of humor. Even as a young man, his veneer was that of a cantankerous cynic. Miller was taller, with long brown hair and a geologist's beard. Gregarious, Dan Miller's idea of fun included planting M 80 firecrackers in cow pies and throwing wild parties for the geology faculty.

Eventually, Miller left to teach at Colgate, but by then Hoblitt had made up his mind that geology was for him. The collegiality between geology students and their teachers was what first attracted Hoblitt to the field. As Miller explained, academic pretensions are lost on field trips the first time a world-famous professor steps behind a bush with a roll of toilet paper. But what committed Hoblitt to the science was the combination of working outdoors at physically demanding assignments and working on problems that generated intellectual sweat. To him, a field geologist was Sherlock Holmes with a backpack.

But derailing his academic path would probably cost Hoblitt his Army ROTC academic deferment, and send him to Vietnam as a second lieutenant. He decided to take the risk, and if he made it through the war he would return to college and pursue an advanced degree in geology. Then, just as Hoblitt switched majors, the Vietnam War started to wind down. The military suddenly had more second lieutenants than they knew what to do with, so they offered the new geology grad student a slot in the reserves.

Hoblitt ripped through the academic requirements and was beginning work on a doctoral thesis when Dan Miller reappeared to change Hoblitt's life again. Miller had landed a job with the U.S. Geological Survey. He was now working for the two scientists who had practically invented the field of volcano hazard mapping. The three-man group had been deciphering the histories of some of the most threatening volcanoes in the United States, those of the Pacific's Cascade mountain range.

Lately, the group had been looking for a paleomagnetic specialist like, well, Rick Hoblitt. They needed someone who could read the magnetic stories embedded in volcanic rocks. (A volcanic rock's magnetic properties are sealed as the fresh rock cools. If, for example, all the rocks in a deposit give the same magnetic direction, it often means that the deposit arrived in place at high temperatures. This would indicate the deposit was a pyroclastic flow. If, however, the rocks in a deposit had random magnetic orientations, that generally meant they were relatively cool when they tumbled into place, probably as part of a mud flow.) Hoblitt jumped at the part-time job opportunity. He turned his Survey work on St. Helens into a dissertation and turned that into a full-time position with the Survey. He became the fourth member of the Volcano Hazard Mapping group.

Hoblitt did not know it when he was hired, but he had entered the world of volcanology at a pivotal moment. In the twentieth century, many sciences—from nuclear physics to the biology of the nucleus—were advancing at warp speed. Volcanology, however, stood as an exception.

Observation remained the heart of volcano science. Even today, the most generous remark one geologist can make of another is that he (and generally it is a he) is "a great observer." While it may seem limited, descriptive science has been the backbone of many fields—astronomy, for example. But in most other fields of research, observation is only the first step in the scientific process. Generally, it is followed by attempts to explain the observation. Those ideas are tested in experiments. The explanations that survive testing are tested and refined again and again until they reveal a powerful truth about nature that can be used to make predictions.

Unfortunately, volcano scientists rarely had the luxury of checking their theories with repeated experiments. Volcanoes were often undependable, remote, and hostile laboratories. These constraints had hobbled the science for centuries. To make matters worse, one mountain would erupt in a style completely different from a neighboring volcano's. Even a single volcano such as St. Helens could erupt in different styles at different times—quietly pumping out thick lava one time and then blowing its top the next. So when a volcano started kicking up, geologists often knew little about what they were up against.

When Rick Hoblitt entered the field, volcano scientists had started to break free from their limitations in two ways. First, they were adapting research tools from other sciences to monitor what a volcano was doing just prior to and during an eruption. These devices were tested most intensively at America's premier volcano lab, the Survey's Hawaiian Volcano Observatory. HVO, as it was known, had two active volcanoes: Mauna Loa and Kilauea. Together, the two volcanoes erupted so often that they gave scientists almost constant opportunities to test equipment and refine theories.

A second entirely different approach had been opened by Hoblitt's new bosses, Dwight "Rocky" Crandell and Donal (no second *d*) Mullineaux, who operated out of the Survey offices near Denver. Their work expanded observation by exposing a volcano's past in unprecedented detail.

When a volcano erupts, no matter what size or style the eruption, the eruption is recorded on the landscape. In explosive eruptions, the peaks of volcanoes can burst like grenades, scattering rock and debris over a wide area. Sometimes eruptions are less violent and extrude thick, pasty lava. Sometimes an eruption can begin with an explosion and end with the building of a lava dome. Other volcanoes can have small eruptions, but trigger major flooding as they melt their cloaks of ice and snow. No matter how they erupt, they leave a record of their eruptions across the countryside.

Beginning in the mid-1950s, Crandell and Mullineaux had begun piecing those stories together from the fallout of composite volcanoes in the Pacific Northwest. The two scientists generally worked where the layers were exposed by road cuts and river canyons. In these places, they could see the layers of eruptive debris piled one atop another. The real challenge of the job was distinguishing one eruptive layer from those that sandwiched it.

To decipher those stories, Crandell and Mullineaux took existing geological techniques from other fields and, for the first time, used them to tease apart one eruption from another. One exploited the fact that the ash from any single eruption was discovered to be almost as unique as a fingerprint. In one sense, volcanoes are the chimney pipes of deep furnaces cooking liquid rock. This white-hot material rises because it is pressurized and more buoyant than the overlying rock. As it rises, it cools, causing the cooking environment to change. Gases escape, temperatures drop, water is mixed and separated, crystals form, and so on. These changing conditions mean that by the time it leaves the vent, it contains a particular combination of mineral crystals that are unique to that eruption. A pioneering volcanologist in Denver named Ray Wilcox had taught Mullineaux how to tell one ash from another by identifying the crystal fingerprints. Mullineaux refined the technique, so that with the simple equipment that he could carry in his backpack, he could distinguish one ash from another while in the field.

The second tool was also one that was being broadly applied in other fields, carbon dating. Crandell and Mullineaux would find chips of burned wood in layers of ash and old soils, and by using carbon dating, they had a second way of separating one eruption from another as well as a way for dating the eruption.

While Mullineaux became an expert in reading the pyrotechnics of an eruption, Crandell worked on deciphering an underappreciated aspect of volcanoes, lahars. On some volcanoes, snow and ice can melt rapidly during an explosive eruption, which can produce a lahar, often called a mudflow.

Sometimes, a volcano's rock can be cooked underground by steaming acid, which can make large portions of a volcano so rotten that it essentially falls apart. The collapsing mountain also creates a mudflow. Crandell had found just such a situation while preparing geological maps of Mount Rainier, the impressive snowcapped volcano that looms over Seattle. Crandell discovered that fifty-seven hundred years ago, a substantial portion of Mount Rainier had given way. This landslide mixed with glacial ice and stream water and became one of the biggest mudflows ever discovered. Known as the Osceola Mudflow, it ran from Mount Rainier more than sixty miles to Puget Sound.

By uncovering a volcano's detailed history, Crandell and Mullineaux had begun to allow geologists to look at some Cascade volcanoes as they had never been seen before, not just three-dimensionally as they stood out on the landscape, but in the fourth dimension, through time. In several places, Crandell and Mullineaux had discovered that today's pastoral plains were really the remnants of mud floods two hundred feet deep. They identified landslides that dammed rivers, which then built lakes, which were destroyed in subsequent eruptions. They tracked monumental clouds of ash that had blown out of summits and fell as far away as eastern Canada.

They felt the most important benefit of their work was that by looking at a volcano's past behavior, they were able to identify the unique threats that some volcanoes would pose in the future. Their work was used, for example, by planners to pick safe sites for new constructions.

Geologists had long believed that when it came to eruptions, the past was the key to the future. But generally, they had been limited in how deeply they could look into the past. Now, they were being given a detailed past to study, one that could reach back tens of thousands of years, with a precision never before available. Volcanoes that had little or no historical record of erupting suddenly were seen to have their history exposed, spread around their flanks waiting to be read.

Powerful though the new approach was, it had two weaknesses. First, in some places, it might be difficult to tell how closely one layer of an eruption was related to the layers that sandwiched it. For example, it might be impossible to tell if a pyroclastic flow in one layer happened during the same eruption as the mudflow that overlaid it, or if the mudflow happened months or even decades later. Carbon dating was good, but its margin of error could be frustrating. The second problem was that powerful events, especially blasts and (less vigorous but still deadly) pyroclastic surges, could leave thin deposits that might be missed or eroded away before they were locked in place by a subsequent event. Thus it was possible that Crandell and Mullineaux might underestimate the reach or entirely miss important aspects of eruptions.

While developing the new approach was an achievement, keeping it funded required another act of scientific wizardry. Throughout the 1970s, the Survey's volcano budget was about $1 million a year, and nearly all of that went to the Hawaiian Volcano Observatory.[1] Rocky Crandell, as the leader of the small Denver-based volcano hazard mapping team, had to convince the Survey that the work would be beneficial to society. It was a tough sell. Volcanoes were not seen as much of a threat when they were not erupting, and the odds of an individual volcano in the Cascade Range erupting during our lifetime are not high. Also, the Survey already had a prestigious volcano research program in Hawaii that had been functioning for decades. It was staffed by people from the Survey's prestigious Branch of Field Geochemistry and Petrology. What could a couple of experts on stratigraphy, based in the Survey's Engineering Geology Branch, bring to the party?

Crandell won modest funding from the Survey by showing that volcano hazard mapping could help land-use planners make wiser choices. At Rainier, for example, Crandell argued there is no real reason why a mammoth mudflow could not happen again. Nothing could stop such a flow from sweeping over the five towns that had sprung up on the Osceola flow. Even if reservoirs were drained before such a release, no dam could hold it back. And it could happen without warning. Tens of thousands of people were at risk. Crandell advised in 1971 that no permanent residence be constructed on the valley floors around the volcano.[2] That advice has been ignored.

At every opportunity, Crandell kept lobbying for support for volcanic hazard mapping. In 1975, Crandell and Mullineaux warned that Cascade volcanoes should not be ignored. "The present quiet interlude, which began shortly before 1920," they wrote, "has prevailed at a time when there has been an ever-increasing use of areas near volcanoes for recreation, permanent and seasonal residences, lines of communication and transportation, lumbering, hydroelectric developments, and nuclear power plants. However, it is virtually certain that this quiet period is temporary and will be ended by an eruption, perhaps not this year or this decade but conceivably before the end of the century."[3]

In the late 1960s, after finishing work on Washington's Mount Rainier and California's Mount Lassen, Mullineaux went scouting volcanoes for their next study site. He took a quick look at Mount Shasta in northern California and concluded that it was not a particularly threatening volcano. About this time, Crandell was encouraging a geology graduate student to pursue his doctorate by looking at the record of eruptions at the volcano nearest to his home, Mount St. Helens. The student reported back that the site had seen plenty of activity, and Mullineaux went to assess it for himself. It didn't take Mullineaux long to find layer after layer of debris. "I was just goggle-eyed. There were too many to count," Mullineaux would later say.

Crandell submitted a research proposal to examine St. Helens. His justification for the project was that three hydroelectric dams lay in the path of potential mudflows, and a modern mudflow could cause the dams to fall like dominoes. "More than 40,000 people live on the valley floors downstream from the reservoirs," Crandell wrote in his September 1969 proposal. They were in danger. The estimated cost of the project: $36,700 a year. Approved.

One of the first purchases Crandell made with the funds was a twenty-two-foot Holiday trailer. The trailer became the field headquarters and housing for the Mount St. Helens Volcano Hazard Mapping group.

After a few years of careful study, Crandell and Mullineaux reported the mountain had been quiet for five thousand years when it began its latest round of eruptions, about thirty-five hundred years ago. Shortly after the mountain reawakened, St. Helens pulverized a cubic mile of its edifice in its most massive eruption. In size and chemistry, that eruption was similar to the A.D. 79 eruption of Mount Vesuvius. Ash from the eruption was found near Hudson Bay in northeastern Canada. Spirit Lake began to form shortly after this huge eruption when a mudflow came to rest in a narrow valley at the foot of the mountain, creating a natural dam across what is now the North Fork of the Toutle River.

Crandell and Mullineaux also found that mudflows and debris avalanches had sometimes plugged other valleys and created other lakes around St. Helens. But these other natural dams were not permanent. They gave way when they filled and a small trickle of water began to cut through the dam. The trickle became a river as pressure behind the dams made the debris walls fail catastrophically. The bursting lakes then launched their own floods of mud. These prehistoric floods swept away vast stands of old-growth trees and boulders as big as houses, and miles downstream they left layers of mud tens of feet thick.

After nearly a decade of hiking and digging and analyzing and dating and mapping, Crandell and Mullineaux, and their young protégés Dan Miller and Rick Hoblitt, came to see St. Helens as no one else ever had. In 1978, they published a paper identifying St. Helens as the youngest and most violent of all the Cascade volcanoes. While the volcano's throat was more than thirty-six thousand years old, much of what was visible as Mount St. Helens had been built up only in the last twenty-five hundred years. In other words, St. Helens was a geologic infant still experiencing growth spurts. It had exploded dozens of times, and the explosions were often followed by periods of lava extrusion that rebuilt the mountain. By comparison, nearby Mount Adams had never had an explosive eruption in historical times. Crandell and Mullineaux warned, in that 1978 report that became known by its binding as the Blue Book, that St. Helens was "an especially dangerous volcano."

"Mount St. Helens has been more active and more explosive during the last forty-five hundred years than any other volcano in the conterminous United States," they wrote.

As the geologists were able to determine the dates of the eruptions, they discovered that St. Helens had become somewhat predictable. It had been popping about every one hundred years (The last major eruption had ended in 1857) and by the late 1970s, St. Helens was overdue. In the 1978 Blue Book, the scientists concluded: "The volcano's behavior pattern suggests that the current quiet interval will not last as long as a thousand years; instead an eruption is more likely to occur within the next hundred years, and perhaps even before the end of this century."

The geologists speculated that people living near the volcano would be the first to notice signs of unrest. These early signs would be earthquakes and snow avalanches caused by the quakes or by the mountain swelling from new magma intrusion. And eruptions in the future would be of the same "kinds and scale as those which occurred repeatedly during the last four thousand five hundred years."[4]

In a "Notice of Potential Hazard," which included the Blue Book and a cover letter discussing the hazards, the geologists predicted that if a future eruption of St. Helens followed the volcano's habit, then "the greatest potential danger will exist at or soon after the onset of volcanic activity."

The "Notice" was published in December 1978. When Washington State officials read the Blue Book, they assumed the mountain would explode any day. They anxiously organized a meeting that was held in Olympia, Washington, on January 8, 1979, to prepare for the apocalypse. At that meeting, Survey officials reassured the state employees that the mountain was not currently building steam for an imminent eruption, but it would be wise to prepare.

When the panic subsided in Washington State, not much happened. The state's Department of Emergency Services was a political backwater, with a small staff and a minuscule budget. State officials were not going to waste their resources to prepare for an eruption that might never come or at least not during the current administration.

The most common reaction to the Mount St. Helens forecast was reflected in a local newspaper that wrote: "One sure way for geologists to make the news is to predict one of the Cascade Range volcanoes is going to erupt again."

Not much happened within the Survey either to prepare for an eruption of America's most dangerous volcano. Those geologists who saw themselves as the real volcano experts in the Survey, those who were working with advanced equipment on real eruptions in Hawaii, criticized the warning as stating the obvious. Said one: "It wasn't rocket science." Higher up the Survey's food chain, other matters were more pressing. National calamities were raw meat for the Survey, but St. Helens was only a disaster in waiting. So the Survey had little time or money for it. One Survey official, Robert Tilling, who oversaw volcano programs at the time, wrote later: "Ironically, the Survey itself failed to respond institutionally to a long-term forecast made by its own scientists. Because of budgetary constraints and other (at the time) higher-priority programmatic commitments, Survey officials (myself included) made no decision to begin baseline monitoring studies at Mount St. Helens."[5]

Inside the Survey, it was also becoming apparent that the agency lacked the resources to confront any volcano emergency. In fact, an internal Survey memo written in 1980 detailed "how poorly prepared the Survey would be for an eruption at any of the [Cascade volcanoes]."

If there were some who feared an eruption at St. Helens, Rick Hoblitt was not one of them. Like other geologists who had long worked on a volcano, Hoblitt could envision the many different geological shows St. Helens was capable of staging. Based on his work, he put a lot of detail on those imaginary eruptions. Like other young scientists in his field, Hoblitt was in love with volcanoes, their power, their beauty, and the geological mysteries they produced. It wasn't just St. Helens. Hoblitt had read descriptions of eruptions at other volcanoes. One of his other favorites was a killer called Mount Lamington. Hoblitt badly wanted to see this elusive beast, and he especially wanted to see his volcano erupt. In his office locker in Denver, he kept a "hope chest" of cameras, asbestos gloves, thermocouples, and other eruption tools just in case he got lucky.

So, Hoblitt went about his work with a gnawing concern that he could spend his entire career on a mountain that might never erupt. For most of the year, he worked in a lab in Denver adding more detail to each of the St. Helens eruptions. Then, each summer, he and his wife would return for the field season to their trailer parked beside Spirit Lake, where he would lug his frustrations and his balky rock drill up the mountain and down river valleys.

Often enough, there were nights like the one in August of 1979 when he pitched his tent and fried some meat near Floating Island. After Hoblitt and his summer field assistant had finished their meal, but before the sun was completely gone, they clambered over the perilous black river of rock to one of the forested islands and had a drink. On the island, he would do something that was becoming a ritual. He would raise his tin cup of booze to the mountain and yell, "Why don't you erupt, you son of a bitch?"

Soon, it did. And the Floating Island flow was obliterated.

CHAPTER 2

Disbelief

Across the quiet winter forests of Mount St. Helens, trees shuddered one after another. Snow dropped from their limbs in a wave that rippled down the mountain, down to the Toutle River valley and across Spirit Lake. The motion of the earth began rolling the one seismometer on St. Helens's west flank at 3:47 P.M., Thursday, March 20, 1980. A U.S. Forest Service avalanche watcher stationed at Spirit Lake felt as if she were suspended from a swing, bouncing in all directions. The earthquake was gentle and felt only on the north side of the mountain. Near the peak of the mountain, the ground movement released a broad avalanche that surged down a chute that two climbers had just abandoned. The avalanche narrowly missed a herd of snowmobiles, and sprayed into the unoccupied Timberline parking lot. By luck alone, the first stirring of Mount St. Helens in more than a century injured no one.

The avalanche watcher radioed news of the quake to Forest Service headquarters in Vancouver, Washington, a nearby community across the Columbia River from Portland, Oregon. Because of the recent eruption prediction, a Forest Service official called the U.S. Geological Survey in Denver. Donal Mullineaux took the call and, in turn, called the National Earthquake Information Service in nearby Golden, Colorado. Seismologists there said the quake registered 4.2 on the Richter scale.[1] The information Mullineaux found most interesting was that EIS scientists had located the epicenter of the earthquake fifteen miles north of St. Helens. Mullineaux relayed the information to the Forest Service and added there was little to worry about. Earthquakes in the Pacific Northwest are not uncommon, and this one, located at such a distance from the mountain, was not likely to be connected to the volcano's plumbing system.

The quake had also tripped a seismic net operated by scientists at the University of Washington. In 1972, in a burst of foresight, the Survey had installed seismic stations on Rainier, Baker, and St. Helens. Their sole purpose was to monitor for volcanic quakes, which Crandell and Mullineaux had predicted would be the first sign of an awakening volcano. Eventually, the responsibility for monitoring the instruments was turned over to a lab at the University of Washington run by Steve Malone, professor of seismology, and his colleague, from the U.S. Geological Survey, Craig Weaver, a former student of Malone's. Theirs was the instrument on St. Helens's west flank.

On this day, the box that encased that seismometer rolled. A pendulum inside the box remained stable, but the relative change between the box and the pendulum generated a small electrical pulse. The jolt was translated into a radio signal that bounced off a set of repeaters back to Seattle, where a stylus scratched out a wavy line on a drum recorder in the basement of the UW's brick and ivy-covered geophysics building. In fact, a row of seismographs began twitching as the quake rippled across the Northwest. Weaver looked at the markings, ran up two floors from the basement seismic room, barged into Malone's office, and announced, "We've just had a big earthquake down at Mount Hood," a volcano to the south of St. Helens.

The two seismologists raced back down to the basement lab. With only a quick glance at the tracings, it was obvious that the quake was larger than anything else they had seen in the region. But Malone, like Mullineaux, wanted to know where the quake had originated. At the time, calculating the origin of the quake was a complex job. Yes, the St. Helens instrument had jumped the most, but perhaps it was recording a signal from nearby Mount Hood, just across the Columbia River from St. Helens, which didn't yet have its own instrument.

While the two scientists traded speculation over the location, a young assistant began working the data. As usual, the quake had registered on different seismometers at different times. It was like a wave reaching an irregular shoreline. By taking the arrival time and intensity from each location and triangulating backward, intersecting lines would cross at the quake's origin. Simple, in theory.

Seismologists, then and now, walk a thin line between science and art. The fundamental problem in pinpointing a location is that there are never enough instruments to blanket an area, or enough time to analyze the data completely; nor, most important, is there enough understanding of the earth through which the seismic wave travels. Seismic energy travels at different speeds depending on the medium it is moving through. A dense medium such as a solid piece of granite will transmit the wave quickly. But a geological junk pile like St. Helens was a challenge.

Solving the problem, in those pre-PC days, meant punching arrival times from the different stations onto computer cards. After that was done, the stack of cards had to be run through a batch sorting device that took even more time. The job could be done in an hour by hand, but it seemed more scientific if it was done at the university's computer center. It still took an hour. By then, the Washington University group of scientists and students—which by now included other curious geologists working in the building—had the results. The quake had originated directly under St. Helens.

Hours later, one of the people in the lab, Elliot Endo, a former Survey technician now studying at UW, noticed something else in the data. The earthquake didn't completely die. Normal earthquakes, tectonic quakes, are generally triggered when two great chunks of the earth skip past one another. Like two pieces of sandpaper, rock masses don't slide easily. They built up tremendous tension, and suddenly, they lurch. The jolt generates a seismic wave that rings like a bell, loud at first but then dying out as the energy from the quake subsides. On a recorder, these quakes look like a Christmas tree, swinging wide initially but eventually returning to rest at the center again.

That is not what Endo saw. Initially, the styluses had swung in big arcs and tapered off rapidly. But the styluses never completely steadied at dead center. The machines kept recording what could be a series of small quakes. If they were quakes, they were so small that they were certainly not felt even at the mountain. Endo said he had seen a similar pattern of quakes while working at the Hawaiian Volcano Observatory. The pattern, said he, was characteristic of a volcanic process. It could mean magma was pushing its way through old, cold rock.

Perhaps, said Malone, but the possibility that these were volcanic quakes was remote. It was more likely a main-shock/aftershock sequence that was common in the Pacific Northwest.

Yet the quakes continued, which made the case stronger that these were volcanic quakes. Malone wanted a lot more data before he would say anything publicly. The quake could have been large enough that it was generating aftershocks. Malone decided to increase the coverage of the area. The group had only one seismometer on the mountain itself. The next closest instrument was thirty miles away. Coincidentally, Malone had that day received four new seismic machines for a geothermal experiment, and he decided to put them all near St. Helens.

The next day, Friday morning, March 21, Malone and Endo drove more than 125 miles south to St. Helens and planted two recorders on the north side of the volcano. Weaver and another student planted two on the south side. Most of the recorders were near the mountain, but one was planted twenty-two miles away.[2] Malone thought that if the quakes became more intense, which is a characteristic of a volcano on the move, then the nearby stations would be overwhelmed with signals. Distance, on the other hand, would provide a natural filter. Malone selected the sites with an eruption in mind. One recorder was placed on the slope of Coldwater Ridge, about six miles across from the valley. He thought it was far enough and high enough on the ridge so it was not likely to be destroyed in any eruption.

Back in Seattle on Saturday morning, Malone could see that the quakes were not fading as they would if they were a main-shock/aftershock sequence. Overnight, they had even picked up a bit. It was becoming hard to say these were only aftershocks. The data indicated that the earthquakes, most of which were too small to be felt, were moving under St. Helens in swarms.

Malone called the Forest Service and told them that quakes were continuing at St. Helens. The Forest Service ranger was concerned about avalanches and recommended closing the mountain.

Throughout Saturday and then Sunday, Malone's recorders were constantly twitching. Something was rattling the mountain. It just didn't seem possible that this was volcanic activity.

"I'll admit it right out," said Malone years later. "We just didn't believe it was volcanic. That sort of thing doesn't happen to you. We were always trying to think, 'Is this really the case?'"

Disbelief is often the first response to a volcano's awakening. For people living within the reach of a volcano, it takes a powerful effort of imagination to grasp that the neighborhood mountain or hill or valley or lake or harbor or, in one case, cornfield is actually the site of a volcano, a force of nature that could actually rise up and transform the landscape in an instant. This same predisposition to disbelief is also true of scientists.

Malone decided he would stop wrestling with the problem Monday morning. He told a colleague that if the quakes continued into Monday, it would be time to panic. Early Monday, he rode his bike to the campus and at 7 A.M. looked at the overnight readings. It was time to panic. "I was definitely in way over my head," said Malone. He picked up the phone and called Rocky Crandell in Denver.

"Don't worry," said Crandell, who had been told of Thursday's quake by Mullineaux. "That earthquake is something like twenty kilometers away."

"No way," said Malone. "It's right under the mountain."

Malone told Crandell that the activity had been building since Thursday with a 4.0 on Saturday and possibly another 4.0 this morning. Those were big quakes, and with quakes continuing, it meant they were being driven by a lot of energy. Malone hadn't had time to analyze the latest data completely and said he would call Crandell back when that was done.

As Crandell well knew, all the literature on restless volcanoes added up to one thing: they were unpredictable. Any newly rumbling mountain could rush to eruption with as little warning as a few hours. Or they could tremble and spout small eruptions for years before exploding. Or, as Crandell and a lot of Survey volcano scientists could not forget, restless volcanoes could huff and puff and then do nothing at all.

Almost exactly five years earlier, another Cascade volcano had begun showing signs of life. On March 10, 1975, a dark cloud was ejected from the top of Mount Baker. It was another snow-clad volcano 180 miles north of St. Helens, just fifteen miles south of the Canadian border. Geologists flew over the summit and saw that it had opened new steam vents, which indicated a new source of heat. They also saw that the summit glacier had been deformed, perhaps by a quake. Monitoring the activity over the next three weeks, scientists watched the amount of heat discharging from the mountain rise tenfold. High levels of hydrogen sulfide, a gas that was thought to be associated with magma, were detected. Geologists became concerned that even a small eruption would unleash a large mudflow that could inundate campgrounds, top a dam, and flood a broad region downstream.

Based on the Survey analysis, officials made some tough calls. The Forest Service closed mountain campsites and the popular Baker Lake Recreation area. The power company did not fill its dam that spring so the reservoir could contain the largest possible mudflow. The actions were not without consequences. The dam drawdown cost the utility an estimated $775,000 in power sales. Tourists avoided Mount Baker. Restaurant owners, innkeepers, and wilderness outfitters complained that the Survey was overreacting. Business fell in the region by 25 percent.[3] But through the tourist season, as opposition grew loud and angry, the scientists held their ground. Months passed with no eruption. On April 6, 1976, more than a year after the initial event, the Survey issued a reassessment that concluded "there [is] now no clear evidence of forthcoming eruption."[4]

Criticism of the Survey was bitter and enduring. And the criticism wasn't all from outsiders. Overall, the Survey response was so disjointed that it caused a senior Survey scientist, Bob Christiansen, to write a memo in February 1980, just one month before St. Helens began quaking, warning that the Survey was "poorly prepared" for an eruption of any Cascade volcano.[5]

Rocky Crandell, who had been part of the Survey's Baker response, recalled the long and stressful experience as he now began organizing the response to the activity at St. Helens.

After talking to Malone, Crandell called Bob Tilling at the Survey headquarters in Reston, Virginia. Tilling was chief of the Office of Geochemistry and Geophysics, which put him in charge of the tiny Volcano Hazards Program. But Tilling was out to lunch so Crandell next alerted the public affairs office to be prepared to issue a formal hazard notice. He asked a Survey team in Seattle to conduct an aerial inspection of St. Helens. At eleven-fifteen, Tilling returned Crandell's call. In the phone log that he started that morning, Crandell wrote in his careful hand: "Called by Tilling who asked that one of us be Survey spokesperson." Crandell, who was the leader of the Denver group, told Tilling that it would be Don Mullineaux, who would probably leave for Vancouver the next day.

At the Survey headquarters in Reston, Bob Tilling hung up the phone and walked down the hall to brief the chief geologist and ask a question. Tilling was new to upper management. The Survey had a long-standing practice of rotating scientists into management positions, generally for a few years. It was thought that only scientists knew how to manage other scientists, which generally meant getting the Survey scientists as much money as could be wrung from Congress and standing out of the way. Tilling had recently been promoted to headquarters after serving as scientist-in-charge at the Hawaiian Volcano Observatory (HVO). He knew that taking the pulse of an active volcano could be a demanding and even dangerous job. He carried a scar on his back from a chunk of lava that had rained on him when he got too close to one of the Kilauea eruptions.

Tilling also knew that monitoring a volcano could be expensive. For St. Helens, personnel could be temporarily reassigned, but they would need to be transported, housed, fed, and most important, they would need a helicopter to move them around the wilderness. So, after briefing Dallas Peck, the chief geologist, Tilling asked him how large the contingency fund was for such events. Peck asked in turn, "What contingency fund?" There was no extra money.

Tilling walked back to his office, phoned Crandell again, and told him not to worry about money. He said to charge everything to a special account and gave Crandell the number. It was a small, flexible account given Tilling for office supplies and small bonuses. Tilling was gambling. If the volcano quieted soon (the most likely course), little would be spent. If it blew big (an extremely remote probability), no one would question spending the money. The worst possibility was that the activity would limp along for weeks or months, consuming bags of money, and never erupt. In that case, he would have to explain why he needed a million dollars' worth of paper clips.

The day after Malone's call to Crandell, the UW seismologist's instruments were practically jumping. Seismicity was ramping up, both in number and intensity of quakes. On most machines, Malone had difficulty reading one earthquake from another. About all he could tell Crandell was that there were "many magnitude 4.0-plus events." He had never seen this much energy.

On Tuesday night, Don Mullineaux flew to Portland and the next morning walked into U.S. Forest Service headquarters in Vancouver.

The Forest Service had direct control over thousands of acres on and around St. Helens. But that control was complicated by a checkerboard pattern of land holdings on the mountain with the U.S. Forest Service responsible for about half the region and the rest held by logging companies and the state.

The Gifford Pinchot National Forest headquarters building is a two-story blockhouse—gray concrete slabs with slits for windows. If the weather was clear, the back windows provided a good view of the snow-cap on St. Helens, forty-five miles to the northeast. But this was March in the Pacific Northwest, so the weather was rarely clear. When Mullineaux walked inside the building, he found the Forest Service efficiently preparing for disaster. The Survey may have been unprepared to deal with a natural calamity, but the Forest Service had no such handicap. It had decades of experience with catastrophic forest fires.

Mullineaux had barely exchanged morning pleasantries when he was ushered into a small, first-floor conference room packed with about thirty-five people. Following their general disaster script, the Forest Service had gathered all the "stakeholders" for a briefing on the current situation and its potential. Sitting around the table and against the walls were county sheriffs and state law-enforcement personnel, representatives from the state Department of Emergency Services, local reporters, dam operators from Pacific Power and Light, county politicians, representatives from the U.S. Corps of Engineers, and managers for the big timber owners, Weyerhaeuser and Burlington-Northern.[6] The Forest Service gave a brief update on the earthquakes and avalanches, then introduced the geologist.

Mullineaux, who had just turned fifty-five, had wavy dark hair, dark-rimmed glasses and a wardrobe that ran from casual to camping attire. He began his part of the briefing by saying that in his line of work the past was the key to the future. Mullineaux said St. Helens had erupted in the past and could be expected to erupt again. It was a type of volcano that can extrude thick, slow-moving lava, but that such flows generally occurred after explosive activity. An eruption at St. Helens, he said, could include all the tricks a composite volcano can pull. It could blast enormous quantities of pulverized rock, which could blanket downwind areas under several inches of ash. Even without an eruption, a warming volcano could melt snow and ice and launch mudflows that could clear forests, carry boulders, and trigger flooding in the mountain's drainages. And perhaps most dangerous of all, the volcano was capable of pumping out pyroclastic flows, searing winds of gas and ash and pumice that can travel at more than a hundred miles per hour and snuff out all the life in an entire forest. This is what it had done in the past, and that was what it was capable of doing in the future.

Mullineaux said his most immediate concern was the danger of flooding. The mountain was heavy with late-winter snows. Even a minor eruption could quickly convert the snowpack into water or fast-moving mud. Without warning, one hundred thousand acre feet of liquid could come rushing down the side of the mountain. It would be as if a dam collapsed. And if that happened on the south or east side of the mountain, the sudden release of so much liquid could cause the collapse of one or all three hydroelectric dams on the Lewis River. Power company officials said the dam closest to the mountain, the Swift Creek Dam, could contain such a release.[7]

Would St. Helens erupt or not? Mullineaux was asked. "We cannot predict an eruption," he replied. The answer took the group by surprise. A burly man from the state forestry service asked, "You mean to tell us that we as a nation can send a man to the moon and you can't predict if a volcano will erupt or not?" Yes, that's correct.

Uncertainty could get to be expensive, Mullineaux was warned. If St. Helens became more ominous, the Forest Service would shut down logging, ban fishing and camping around the mountain, and close the area even to those who owned cabins on Spirit Lake. Surely, with so much at stake, scientists could give better guidance than "we don't know." Mullineaux shaped his answer in many ways, often giving probabilities, but the answer was essentially the same: there is no way to predict what a volcano will do. Six hours later when the meeting ended, Mullineaux walked out of the room aware of how high emotions could run. It was going to be a tense time even if nothing happened. It would be tense *especially* if nothing happened.

People in the room had one of two reactions to Mullineaux's warnings. A few agreed with a senior Forest Service official who left the meeting thinking, "This could really happen." But most people did not believe the mountain on which they, their parents, and in some cases their grandparents had hunted, hiked, and worked was really dangerous. It was a mountain. A mountain! And even if it did erupt, they didn't believe it would be a major problem.

"I suppose that is not an unusual phenomenon," said one of those at the meeting, "when you consider how you would probably react if a total stranger showed up on your front doorstep and announced that you had an active volcano in your backyard. That's just a little hard to swallow."[8]

Even the few in the room who believed an eruption was possible didn't envision the type of eruption Mullineaux talked about.

"Mullineaux was more convinced that something was going to happen than the rest of us sitting around the table," said one of those at the meeting, Van Youngquist, a dairy farmer and a county commissioner "We were there in awe and disbelief. I'd been to see Hawaiian volcanoes, and that's what I was thinking about. I was thinking about fire and lava flowing."

After the meeting, the forest supervisor invited Don Mullineaux to work out of his offices. The Forest Service had disaster plans ready. They practiced responses yearly. And they had logistical support in reserve. The agency also had an existing communications network around St. Helens, experience in dealing with the media in crisis situations, and employees accustomed to working for long periods under stress. The Forest Service could also provide air coverage, and in emergencies one forest supervisor could draft manpower from other national forests. And they were used to taking the heat for unpopular decisions, such as forest closings.[9]

Most important, the Forest Service was a decisive agency with a military-like chain-of-command, broad legal authority, and the habit of making tough decisions quickly. It was quite a contrast to the Survey, where individualistic scientists tended to stake out their positions, fight for them against the tide if need be, and relent only after being buried by overwhelming evidence.

About the only thing the Forest Service lacked was expertise in volcanoes. And that is why the forest boss wanted Mullineaux just outside his door.

The advantages were obvious and Mullineaux quickly accepted. It put Mullineaux in a role that played to his strengths and established the standard for Survey scientists responding to any volcano crisis in the future. He would provide expert advice but would not make safety or economic decisions. Those would be made by the people hired or elected to make those calls. Mullineaux would advise. This was one of the few points geologists had already thought through. In a paper written in April of 1979, well after the Mount Baker experience, the Denver team wrote, "We are not . . . attempting to determine the levels of risk that are acceptable or unacceptable; this is largely a socioeconomic problem to which we can contribute only one part of the data necessary to make decisions."[10]

Many decision makers depending on the expertise of the Survey found its playing the role of an adviser to be a cop-out. One social scientist who monitored the hazard response in the following weeks wrote that the Survey "attempted to avoid decision making."[11] Many found just getting clear advice from the Survey was not easy. Mullineaux was a careful scientist and spoke like one; nothing was certain, everything was a matter of probabilities. It led Sheriff Les Nelson to say in frustration, "Trying to pin down a geologist [is] like trying to corner a rat in a rain barrel."

To provide effective advice required that the adviser be credible. On March 26, in that small conference room talking about mudflows, pyroclastic flows, and ash clouds, Don Mullineaux didn't really have much credibility.

The little objective proof available at the time that the volcano was becoming dangerous was in the seismic record. Most of the earthquakes were so small that they could not be felt, and those that could were so shallow that they could not be felt much beyond the north slope of the mountain. So, believing the mountain might be building toward an eruption required faith in technical data as well as in the scientists interpreting the records over a hundred miles away in Seattle.

Why should anyone believe this government scientist? He didn't emphasize it, but Mullineaux's expertise was in dormant volcanoes. He had never attended a volcano from its first quiverings to an eruption. In fact, only two or three people in the entire U.S. Geological Survey had any significant experience on explosive volcanoes. The great bulk of the Survey's work had been in Hawaii, on volcanoes that in the memory of all living geologists had produced only slowly rolling rivers of lava, the kind of eruptions Van Youngquist had imagined. St. Helens, on the other hand, was a mountain-size bomb. The Hawaiian and Washington volcanoes were about as different from each other as nature could manufacture.

One thing Mullineaux did know was what St. Helens had done in the past. His job now, as he saw it, was to translate the volcano's past into projections of risk. Then, if people were to be kept out of harm's way and economic losses kept to a minimum, he had to sell those visions to a few utility executives, a couple of county sheriffs, some timber company owners, a handful of Forest Service officials, one cranky governor, some county politicians, assorted sightseers and daredevils, dozens of homeowners, hundreds of loggers, and more reporters than he liked to think about.

Getting people to believe him meant that he had to win their confidence. And at St. Helens, establishing credibility was a problem from the beginning. Following news reports of the Forest Service meeting, people who actually lived next to the mountain sneered at Mullineaux's concerns. "I haven't felt nothing," said sixty-seven-year-old Stanley Lee, who ran the general store in Kid Valley, just a few miles from the mountain. "It's just a crock cooked up by the federal forestry service or them environmentalists to delay a big development of the Spirit Lake recreation areas."

Rick Hoblitt was a believer. When Mullineaux called and asked him to come to St. Helens, Hoblitt broke into his "hope chest" of volcano instruments and was on the plane the same day.

The morning after Mullineaux's briefing and exactly a week after the first quake had rocked the mountain, the director of the Survey sent word from Reston to Vancouver that the Survey would issue a "hazard watch" by 8 A.M. PST. A watch was the second level of the Survey's three-level warning scheme. Survey officials in Reston were coming to believe that a higher warning was justified because earthquakes were very energetic and continuing near or under the volcano. Just the night before, the Seattle group had recorded the one hundredth quake of magnitude 3.5 or greater.

Early that Thursday morning, Mullineaux dispatched Hoblitt to check conditions on the mountain. Mullineaux then went to the Forest Service headquarters where he got word about the hazard alert ordered by Reston. He began notifying more than thirty state, local, and federal agencies.[12]

The mountain had been cloaked by heavy clouds for days, but at 11:20 A.M., an observer with the Army National Guard flying reconnaissance over the mountain reported seeing a hole in the clouds and a gray streak (probably ash) in the ice cap near the summit. Then the clouds moved in and hid the summit before the plane could circle back.

Shortly after noon, Forest Ranger Ed Osmond, who had been assigned to write the agency's playbook, was ready to start writing. As he rolled a sheet of paper into his typewriter and began the first paragraph, St. Helens erupted.

Because of cloud cover no one on the ground actually saw the first eruption. But it was heard. A sound like a sonic boom cracked across the forest at 12:36 P.M. A traffic reporter from a Portland radio station, flying above the clouds near the mountain, was the first to announce it, excitedly telling his listeners, including those in the Forest Service building, what he was seeing: "There is no question at all. Volcanic activity has begun. You can see smoke and ash pouring from the top of the mountain, especially the north side of the mountain."[13] Within minutes, ash began falling on loggers like heavy snow. Weyerhaeuser, one of the biggest logging operators on the mountain, evacuated its crews.

Since cloud cover was so thick, no one could tell immediately if Mullineaux's mud floods were moving toward the Lewis River dams. More than an hour passed before officials were confident the eruption had not turned tons of ice and snow into rushing water. But now the hydroelectric dam operators were true believers and they began to draw down the water levels.

For volcano scientists, the minor eruption signaled an event that could become the opportunity of a lifetime. In the last decade, scientists thought they had made major strides in taking the pulse of threatening volcanoes. Now they saw a chance to test those tools. In Flagstaff, Arizona, a specialist on volcanoes on other planets ran her son to her neighbor's house and said she had to get to St. Helens; could they please watch the child? She didn't know when she would be back. In Hawaii, someone broke into the morning staff meeting at the Hawaiian Volcano Observatory to announce that St. Helens had erupted. (A few days earlier, when it was reported that St. Helens was acting up, the staff meeting disintegrated as the geologists hunted for an atlas because no one knew where St. Helens was.)

At 2:01 P.M., St. Helens was rocked by a 4.7 quake and another small eruption column blasted through the clouds. Airline pilots reported a thick black plume rising to a height of seventeen thousand feet, more than halfway to a jet's cruising altitude. The Federal Aviation Administration then banned all but official flights within a five-mile radius of the volcano. On the ground, Washington State emergency officials advised people within a fifteen-mile radius of the mountain to leave immediately. By midafternoon, dozens of earthquakes stronger than 3.5 were rocking the mountain.

Mullineaux phoned Crandell in Denver and told him the news. After 123 years, St. Helens was on the move. Both men were delighted. They had gambled with their reputations and they had been criticized for making the first official prediction of an eruption. Now, they had been proved correct. Crandell wrote in his journal: *"Our* volcano [is] coming to life, just as we had anticipated." Then, the two men, who knew better than others what the volcano was capable of doing, sobered. Crandell began packing for Vancouver. Dan Miller was already on his way.

Reporters raced to the mountain. A Seattle television station called Steve Malone and offered a helicopter ride to St. Helens in exchange for an interview on the mountain. Malone was too busy but he asked Survey scientist Dave Johnston if he wanted to go. A few days earlier Johnston had been in Seattle attending a scientific meeting when he heard about the quakes at St. Helens and drove to the UW, where he had once been a student. Malone had drafted Johnston into his gang of seismology watchers. Now, Johnston jumped at the chance to see the mountain, and within a hour he was flying under the clouds up the north slope. Johnston saw the chute an avalanche had cut down the mountain's north face the week before. But the summit was hidden. The helicopter landed in the Timberline parking lot, on the volcano's north slope.

The news crew, bundled in puffy parkas, set up the camera with the mountain in the background. Johnston, dressed in a thin blue Pendleton, a turtleneck, a knit cap, and well-used boots, seemed generally unaffected by the wind and the blowing snow.

"[The mountain] is heating up," he said. "Magma is rising. It looks like there's a very good chance there will be an eruption. If there is an explosion, it is possible that very, very hot incandescent debris could come down on all sides. But right now, there's a very great hazard that on this side, the north side, that the glacier is breaking up. That could produce a very large avalanche."

"This is not a good site to be in," he said as he laughed nervously.[14] When he turned to look up at the summit, he had to squint his eyes, grit his teeth, and hold his knit cap on his head. "The fuse is lit, but we don't know how long it is."

While Johnston was being videotaped, another bearded Survey geologist sat across the Timberline parking lot in his government-issued Bronco. Rick Hoblitt, thirty-five, was catching up on his field notes. He had just driven through some of the valleys leading to St. Helens, taking notes and photographing streams and bridges. He drove up the north slope of the mountain making estimates of snow cover.

Hoblitt kept the engine running and began envisioning what would happen if St. Helens erupted—not just a minor throat-clearing like today's pop, but an explosion of blistering gases and red rock. His first concern, like Mullineaux's, was for mudflows. A mudflow would knock down the Douglas fir, drown loggers, skimobilers, and anyone else in its path and then take out bridges, sweep away homes, and destroy at least one dam, maybe two, along the way.

While Johnston and Hoblitt were thinking about what an eruption might do, overhead two Survey geologists from Seattle were circling in a small plane hoping for a break in the clouds to see what the eruption had already done. The two men were specialists in aerial observation, and finally at 4:40 P.M., the clouds parted and they were able to see and photograph how the peak of St. Helens had changed and was continuing to change. The geologists even saw the summit area heaving.[15]

The flying observers saw that the eruption had blown a hole through the snow in St. Helens's summit. In the center of the flat summit, a new crater had been created about 250 feet wide and 150 feet deep. The snow surrounding the new hole was dark, covered with ash.

The blast had smashed the tops of Wishbone and Shoestring Glaciers. In the snow and ice sheets just below the summit, the observers noted an enormous amount of surface cracking and deformation, which indicated the area had been thoroughly shaken. The most interesting cracks in the snow cover were on the north slope. Not far below the summit, the observers saw two long jagged lines, like cracks in a plaster wall. Both fissures ran east-west. One was a mile long and the second was more than three miles long. One of the airborne geologists, Austin Post, speculated that perhaps a plug of lava had risen up into the mountain "or earthquakes have shaken so much that [the] summit is sliding to north, northeast."[16] While the two Survey observers were circling, they radioed that they could also see these cracks open and close.

The careful observers also noted what they didn't see. For example, despite the alterations at the summit and on the north side, there was no obvious damage on the south slope. And there were no large blocks of rock lying in the snow, which indicated that the eruption may have been largely steam, called a phreatic eruption, and not magmatic. Still, even steam eruptions were certainly produced by a new, intense heat source. The other geologist flying with Post, Dave Frank, caught a glimpse of what he thought was a mudflow, but by the time the plane circled back, the clouds had covered the area. They also could not see down to the seven-thousand-foot level, twenty-six hundred feet below the summit, to confirm reports of flowing lava.

With that information in hand, Mullineaux walked into his first press conference at the Forest Service headquarters. Donal Mullineaux took a look at the array of microphones and rows of reporters and was "just appalled." He thought about turning to run, but the Forest Service spokesman—a much larger person than the wiry Mullineaux—was blocking the way. Mullineaux adjusted his glasses and proposed that the television crews turn their cameras off so everyone could just talk. Bad idea. He would say years later that he was uncomfortable doing the daily (sometimes three times a day) press conferences. And, years later, many reporters in the crowd would agree with him. It would be a difficult relationship. Mullineaux expected informed questions, which he rarely got. And following a bad story, he would often spend time he could not really afford to point out errors that were rarely corrected.

Risk communication is a field of science in itself, but it wasn't practiced much at the Survey. The Survey is a science organization where authority is measured by papers published. Those with a wealth of papers were highly prized, and it was assumed everyone else would see their value, too. To Survey officials, Mullineaux was half of the team that had created volcanic hazard analysis, therefore he was the best front man for the USGS at St. Helens. But what was needed was a geologist who could teach reporters about pyroclastic flows and phreatic eruptions and do the most difficult job of all: convey the complexities of a high-consequence, low-probability event. What they got was a rigorous scientist with a reserved bearing, who was slow to reach any conclusion, but once he did he was unshakable. He could not understand the pressures reporters were under. Recalled Mullineaux, "I'd give them facts, but they wanted predictions. To me, they wanted things that scientists could not do."

At his first press conference, Mullineaux told reporters that as grand as the day's eruption had seemed, it was not a big event, certainly not a major explosive eruption like the ones St. Helens had produced so many times in the past. The AP reported that in Mullineaux's view a "full-scale 'pyroclastic' explosion—one which would hurl rock and lava from the mountain—was the least likely possibility. 'A series of small eruptions would be more likely than a large, cataclysmic event,' Mullineaux said."

Not everyone agreed. The next day, a local geology professor named Leonard Palmer was quoted in the newspapers as saying, "You'd have to have your head in the sand not to be worried."

This was precisely the type of public bickering that Mullineaux had hoped to avoid. Mullineaux had feared that other scientists would undermine the Survey's credibility by publicly disagreeing about the threat the volcano posed. The Survey could quickly rein in Johnston, who was one of their own, but they had no control over academic geologists.

Geologists battling could be a risk in itself. Mullineaux had been reading about it in a series of letters that had been running in the *Journal of Volcanology and Geothermal Research* the last few years. The letters were from scientists debating the crisis response to the 1976 eruption of the Soufrière volcano on the Caribbean island of Guadeloupe. In that eruption, the most disastrous effect of the emergency came from the scientists themselves.

After a brief period of intense earthquakes, Guadeloupe's volcano rumbled back to life at 9 A.M. on July 8 when fissures opened on the summit and ash billowed out so thickly that it produced an umbrella over the entire island, blotting out the morning sun. Thousands of people living on the sides of the volcano fled, many clinging to the outside of trucks and cars. When they arrived at the foot of the volcano, they were covered in ash and looked like statues. Shortly, a group of French scientists who had been called to the island concluded that a catastrophic eruption was imminent. That analysis led one island official to say, "We have begun what we think is the countdown. The volcano cannot turn back."[17] The *préfet* [governor] of Guadeloupe ordered seventy-two thousand people evacuated. The news of the impending disaster crushed the local tourist economy. Massive financial aid was required from France to support the relief work.

The burdens on the government, the economy, and the dislocated islanders were exacerbated when a second group of French scientists arrived and announced that their studies of the volcano indicated it would not erupt. Installments of the spat between the two groups were reported daily in the local newspapers. Eventually, an extraordinary panel of international scientists was convened in Paris, which ruled that the danger had passed. After fifteen weeks, people were allowed to return to their businesses and homes.[18]

The postmortem, which Mullineaux had followed in the geology journal, criticized every player, from reporters to politicians to geologists. What was clear to Mullineaux was that the Survey, at least, had to speak with one voice, and as much as he didn't want the job, that voice would be his.

Two days after Johnston and Hoblitt arrived, Rocky Crandell arrived in Vancouver. Walking into the Emergency Coordinating Center, soon to be known only as the ECC, Crandell was stunned by the frenzy. Reports were coming over the radio from the flying observers, typewriters were clicking out updates, the fire boss was issuing orders, and the phones were something else. It seemed to Crandell that as soon as a phone was cradled, it rang.

"Before you get involved with this game," Mullineaux said, nodding at the phones, "the Forest Service and some other agencies have asked for new hazard assessments. Could you prepare a new hazard map for three different sizes of eruptions?"

It was about 8 P.M. then. Crandell found a Forest Service map of the region, locked himself in an office, and spent the night sketching what St. Helens might do. If a volcano's past was the key to its future, then he knew pretty much what St. Helens was capable of. In the Blue Book, Crandell and Mullineaux had assumed "that [future] eruptions will be roughly of the same frequency, kinds and scale as those which occurred repeatedly during the last forty-five hundred years."[19]

Crandell drew the worst possible eruption the volcano could produce: a vertical explosion of ash that would be blown northeast by the winds off the Pacific and drop several inches of ash on towns from here to Montana. He also pictured thousands of tons of snow suddenly melting and flowing like a broad tidal wave down the sides of the volcano. The floods would launch into Swift Dam and possibly overwhelm it, leading to a cascade of floods down the Lewis River valley. The superheated pyroclastic flows he envisioned would blast outward at hurricane speeds. Avalanches would run off St. Helens maybe as far as Spirit Lake. The other two maps Crandell made that night were of smaller eruptions. Crandell finished the maps at 5 A.M. The "worst case" map would become the standard reference for all discussions and decisions about risks to the day of the final paroxysm. It set the worst-case boundary.

In the days that followed, the Denver team, Crandell, Miller, Hoblitt, and Mullineaux, ran on adrenaline. They worked night and day, and then they fell into bed for three or four hours of sleep. The Forest Service now had a plane constantly circling St. Helens, and Miller or Hoblitt would try to take two of the four-hour flights a day. In between, they would be answering phones and briefing on volcanic hazards. One social scientist monitoring the volcano response wrote that many of the Survey scientists worked "up to twenty hours a day for several weeks."[20]

"Every day, I had very long, long lists of things to do," Miller recalled. "There were long lists of calls from the media, other scientists, arrangements, decisions. I always had huge long lists of things I was supposed to do. Many were really important things, things with consequences. I remember falling down and getting four hours' sleep and thinking that I never got half the things accomplished I was supposed to have accomplished. It was a case of feeling an overwhelming sense of responsibility for things that were important, and having nowhere near enough time to do it. It was a frustrating, difficult feeling."

In Vancouver, the mixture of excitement and responsibility drove the Denver team harder than they had ever been driven before. Rocky wrote: "Because of the stress and loss of sleep, my health rapidly went downhill and I found it hard even to think coherently."[21]

A lot of the work these scientists did was simply controlling rumors. People called from all over the West with concerns about pyroclastic flows and toxic gases. People called to confirm if it was true that lava was rolling down the sides of St. Helens and if Mount Rainier had started erupting.[22]

Inside the Forest Service headquarters, the first days after the eruption were filled with meetings, interviews, phone calls, and briefings, and everything was billed as an emergency. Logistics, air traffic control, access to restricted areas, communication problems, evacuations, frightened residents, psychics (who seemed to be in abundance in the Northwest), and reporters all demanded attention.[23] Much of the response they got was being created as the situation developed. Cowlitz County sheriff Les Nelson described the work in this period by saying, "We were building a boat and rowing it at the same time."

After the initial steam explosion, the Forest Service had swung into its fire-fighting mode. An Emergency Coordination Center was established on the second floor of the Vancouver headquarters and staffed twenty-four hours a day. The Forest Service also provided a communications room on the second floor for local and state agencies. Two rooms and six phones were set aside also on the second floor for the Survey. A fleet of twin-engine spotter planes were rotated over St. Helens so, weather permitting, someone would have an eyeball on the mountain twenty-four hours a day. A legendary fire boss, Paul Stenkamp, arrived to direct the troops. And perhaps most important, the Forest Service began lining up fresh replacements.

Everyone was now pushing his way through the Crandell-Mullineaux Blue Book on St. Helens. It became "required reading" in every government agency, industry suite, and newspaper office in the Pacific Northwest. Two hundred copies were sent in the first shipment to Vancouver, and they all quickly disappeared. The paper that had at one time been dismissed was now invaluable. Based on the historical record, power companies knew the largest possible mudflow they would have to contend with. City planners downwind knew how much ash might fall on them. Lumber companies knew which areas were safe and which had been torched by pyroclastics in the past. An independent analysis years later concluded: "Without the Survey data, the warning activities would necessarily have been less authoritative, more speculative, and undoubtedly less effective."[24]

The evacuation of the Forest Service's Pine Creek outpost showed the potential of the Crandell-Mullineaux work. Shortly after the mountain exploded on March 27, the ranger at the Pine Creek station got a call from Ed Osmond at headquarters telling him to "get the hell out of there." Just that morning, Forest Service planners saw that the quickest way out for the Pine Creek staff would have taken them through a zone where there was a high risk of mudflows and pyroclastic flows. So the route that was drawn skirted those areas, and that's the way the personnel were evacuated.[25]

If any other volcano on the U.S. mainland had started to act up, no one would have known how much to lower dams or what areas were unsafe for loggers, or which homes would be threatened by floods. Without having the mountain's history laid open by the Denver team, scientists could have been publicly scrapping over impacts, leaving planners confused and the credibility of the scientists in shreds. Maybe it wasn't rocket science, but to the people who lived and worked around St. Helens it was more valuable: it was the best guidance science could produce.

But the problem with having so much data was that it gave scientists a sense of confidence that they knew what they could expect. Crandell and Mullineaux's work had set the parameters, but in working within those boundaries, they downplayed the possibility of something far larger. By living by the geologist's guiding rule, that the past is the key to the future, they had narrowed their vision to the known history of St. Helens and the confines of the Blue Book.

By the end of March, the volcano was quickly becoming the state's top tourist attraction. The roads were jammed with sightseers, and some of the curious were challenging sheriffs who were trying to hold them back. A county emergency official was quoted in the *Tacoma News Tribune* saying, "People are swarming in from all over, putting their lives in danger. . . . Sunday, when the weather was clear, the road up to the mountain looked like downtown Seattle at rush hour."[26] By April 3, Governor Dixy Lee Ray had declared a state of emergency and the National Guard was called out to reinforce the roadblocks.

Logging companies, too, wanted to get closer to the mountain, and state officials capitulated and allowed three hundred loggers to return to the St. Helens forests.

As always, one of the first jobs for the volcano scientists was to establish an observation post. The Forest Service was flying twenty-four hours a day around the mountain, often with a geologist on board, but the weather frequently hid the summit or blocked observation downslope.

The observation post became Hoblitt's responsibility, and his first job was to pick the location. A spot was needed on the north side since that was the side where the fractures were seen, avalanches had slid, and earthquakes were felt. The Survey scientists wanted the post to be high on a ridge so that the intervening valley would serve as a barrier against anything that might shoot out of the mountain. Mudflows and pyroclastic flows were gravity-driven phenomena and generally followed river valleys and were blocked by high ridges.

Hoblitt chose a spot on an exposed hillside on the north side of the Toutle River valley, about eight miles northwest of the summit. It wasn't ideal. Radio communication was poor from the site and the distance to the mountain made observations difficult. But it was the closest site that could be established until the snow melted and the logging roads opened.

Hoblitt did a quick study of the area, looking for indications that past eruptions had reached the site, which would mean an observer would be vulnerable. He didn't find anything, but he had an uneasy feeling about the site. So he asked Crandell to come out and do his own hazard assessment. Crandell surveyed the site. He, too, failed to find any evidence that the spot had been blasted by pyroclastic flows or topped by mudflows, but he also felt uncomfortable with the site. Before leaving, Crandell said to Hoblitt, "I have a feeling there's going to be a tragedy here."

On March 30, Hoblitt moved into the outpost. The Forest Service had supplied a military surplus tent. The green canvas shelter was large enough for an army cot and a writing table, but it had been well used over the years. It had so many rips that it provided little protection from the snow, rain, and wind. And winter weather around St. Helens was miserable.

Hoblitt would not work there twenty-four hours a day, but he often would work nearly twenty hours a day somewhere. A typical day would begin with a general meeting of scientists in Vancouver. Hoblitt would then work at the Forest Service's command center. Hoblitt could spend much of his day on the phone answering questions from inquisitive reporters and frightened homeowners. Then he would take a seat on one or two observation flights a day. He would also make sure that the observation post, known now as Coldwater Ridge, was staffed and the geologists on duty recorded all activity in the log and on film. Often enough, in the late afternoon, when no one else was available to work at the observation post, Hoblitt would drive up the snowy road to Coldwater Ridge and work it himself.

On one typical day, Hoblitt arrived at the Coldwater Ridge observation post by 6:41 P.M. and began recording what he saw in his thin, brown field notebook:

19:18 Ash fall has ceased.

19:36 Raining, no radio communication, effective observation almost impossible.

19:53 Almost total darkness. Set out coffee can ash sampler with coffee filter paper.

20:20 Pan put under Bronco [for ash].

21:25 Explosions. Earthquakes. Rain continues heavy.

22:27 Quake—pretty strong.

[next morning]

2:20 Summit visible. No rain.

2:22 Weak [eruptive] activity.

2:26 Continues, appears ash rich. Could be moving down north or northeast from summit.

2:53 Mild activity at summit and clouds barely above summit.

6:26 Activity apparent, ash rich plume.

6:48 Emission subsides.

8:47 No summit observations possible since last entry, cloud cover now down to Timberline.

8:50 Raining like hell.

After dark, Hoblitt tried to get some sleep. But the cot was like a seismic sensor. It was pointed directly away from St. Helens, so the quakes bobbed him up and down. By 3 P.M., Hoblitt was relieved and he drove an hour back to Vancouver to get the next observation flight.

Flying with the Forest Service was more comfortable duty. In early April, Miller was in a Forest Service Aero Commander circling just the mountain. St. Helens was, as usual, socked in with clouds just covering the peak of the mountain like a wide-brimmed hat. But Miller soon spotted a white thread of steam wafting above the clouds, and figuring that for the summit, the pilot began circling the spot. Then, in complete silence, the vapor turned black.

"Suddenly," said Miller, "the most incredible, black, nasty-looking cloud came boiling up out of this clear white sea of clouds. It was a phreatic [steam] eruption, and it was a magnificent one. I took at least one hundred pictures of it, and ninety-five of those pictures all looked the same. I remember thinking, 'Oh my God, this is the most incredible thing I've seen in my entire life.'"

Soon, the U.S. Forest Service, which was responsible for air traffic control in the area, had twenty civilian-owned tourist planes stacked up within five miles of the mountain, each waiting for a turn to get a close pass at the steaming, puffing new peak. They were not disappointed. The volcano began chugging like a locomotive with minor eruptions lasting from a minute to forty-five minutes.

Around the mountain, locals were enjoying the show provided by the mountain, the sightseers, and the reporters. A shovel stood outside the Cougar general store with a sign tacked to it that read: EMERGENCY LAVA SHOVEL. Sociologists began interviewing residents and reported inhabitants were "getting a big kick out it."[27] A reporter for a California newspaper asked people near the volcano if they were scared. "Hell, no," said a logger wearing red suspenders. "But did you hear about the herd of forty bigfoot with suitcases that came running down the mountain last night?"[28]

In the village of Cougar (pop. 150) on the southwest flank of St. Helens, reporters often outnumbered the locals in the Wildwood Inn. Sandy Mortensen, who ran the place, could recite each of the television appearances she had made. Volcanoes, she was learning, were good for the tourist trade.

"We've got big herds of elk around here," she told a reporter from a Seattle paper. "We've got terrific fishing. We've got a small gold rush. And now we may have a volcano."[29]

Washington's governor, Dixy Lee Ray, climbed into a state plane and flew to the mountain. Her plane circled for some time before the governor witnessed a small eruption. Still, she was excited by the event: "I've always said I wanted to live long enough to see one of our volcanoes erupt. As volcanic eruptions go, this was probably a minor one, but it was terribly exciting to see a geological event right in our own backyard."[30]

Trying to keep tourists from getting killed, county sheriffs quickly went through their contingency funds and were running on credit.

Maintaining roadblocks on the highways was becoming a joke. The area was laced with logging roads, and soon maps sprouted showing tourists how to skirt the law. A handful of people who snuck around the road-blocks actually climbed the mountain, peered into the crater, and were lucky enough not to get their heads blasted.

Things were just as chaotic in the air. On one afternoon, seventy air-craft loaded with media and sightseers were counted circling the vol-cano. The Forest Service pilot radioed the greatest danger was now dodging other airplanes.

The press was clamoring for news, and crusty Harry Truman, who had refused to evacuate his Spirit Lake lodge, was becoming the poster boy of disbelief and disobedience. The old man defied warnings from sheriffs, rangers, and the mountain. By thumbing his nose at nature, science, and the law, Truman quickly became the most visible reminder that, in the West, fear of the unknown had always been stomped out by gumption.

"I'm gonna stay right here and say, 'You old bastard, I stuck it out fifty-four years and I can stick it out another fifty-four,'" Harry told a reporter from the Associated Press.

Everybody wanted to be as close to the volcano as Truman. The Spirit Lake homeowners were demanding to be let back into the area to retrieve belongings. The logging companies were demanding they be allowed to return to their richest timber sites on the mountain's flanks. The roads to the mountain became jammed with gawkers, including a rocker from L.A. who wanted his new album cover shot with him in the crater. Vendors staked out wide places in the road and sold hot dogs and T-shirts. One deputy spotted a man in a car with Texas plates and said proudly, "Bet you don't have anything like that in Texas."

"No," said the Texan, "but we've got a fire company in Amarillo that'll put it out for you."

CHAPTER 3

The Musketeers

The day after the first eruption, the mountain ejected black columns of steam and ash every twenty minutes. Two days later, observers counted ninety-three different eruptions. While sightseers were awed by the displays, Survey scientists were interested in their color. What made the explosions dark?

St. Helens was probably doing what many volcanologists call throat clearing. As volcanoes warm, especially snow-covered volcanoes like St. Helens, water locked inside the mountain becomes superheated. That is, its temperature rises beyond the boiling point. It would turn to steam but, inside the volcano, the superheated water doesn't have room. To become steam, water needs room to expand. In fact, it needs to expand seventeen-hundred-fold at sea level. If a half gallon of milk flashed to steam in the average refrigerator, it would bust the refrigerator and blow off the kitchen walls. But inside volcanoes, the expansion is contained by the surrounding rock. So pressure builds. Occasionally, the force of the superheated water becomes greater than the rock can contain. Then, a rock cracks and provides an outlet. Instantaneously, steam escapes, often with a roar like jet engines at full throttle. Geologists knew this process was driving the eruptions.

While steam explosions are white, the St. Helens phreatic eruptions were gray, sometimes inky black. The scientists believed that St.

Helens's eruption columns were dark because they contained tons of old rock that was being pulverized by the blasts. But they didn't know if the eruptions also contained something else, fresh rock. If it did, that would mean magma was not only involved but that it was close to the surface. If fresh rock was in those eruption clouds, it would confirm that this wasn't a false alarm but a real magma-driven volcanic emergency. The only way to find out was to get a sample of the ash.

To get a sample, geologists had to get close to the vent. Newly arrived Survey scientist Don Swanson decided it was his job. He went to a hardware store in downtown Vancouver where he bought a yardstick and a soup ladle. He taped the ladle to the end of the yardstick, drove an hour north to a baseball field near the volcano where the Survey's recently leased helicopter was parked, and told the pilot to fly up the side of the erupting mountain. The two men hovered just below the summit while trying to get a sense of the rhythm of the eruptions, and when one blast stopped, the helicopter darted to the mountaintop. In high winds, the former Vietnam pilot held the chopper's skids inches from the lip of the crater. Swanson pushed open his door, leaned out, and began scooping fresh dark ash. If St. Helens had blown then, it would have taken out the helicopter, the pilot, and Swanson. With the samples hauled in, the chopper raced downhill. The ash was analyzed and no new magma was found.[1]

This further strengthened the notion shared by some geologists that magma was not involved. This, they argued, was another simple heating event like Mount Baker. And that would have been the consensus if it were not for the continuing high level of earthquakes shaking St. Helens. But by the end of March, hundreds of quakes were smearing Malone's seismographs daily. That level of seismic activity meant that whatever was shaking St. Helens was driven by enormous energy, and it showed no sign of declining.

The scientists reasoned that if this was a true magma-driven volcanic event, then months earlier, a plug of liquid rock had budded from the main magma reservoir less than five miles beneath the surface. This red-hot gob of rock was more buoyant than the surrounding rock because it was hotter and because it contained dissolved gases. Powered by its relative buoyancy, the plug of magma began to rise through an ancient passageway to the surface. As it climbed, the plug pushed aside, melted, or simply snapped the layers of cold rock that clogged the old route. That process caused the earthquakes that triggered avalanches and twitched Steve Malone's seismometers.

After climbing miles, magma can just run out of energy and grow cold and solid inside the volcano. Or it can flow gently from an opening until its rising energy is spent. In some cases, buoyant magma can even puff up a mountain, then grow cold and solid.

The work that the Denver team had done in uncovering the volcano's history had allowed the Survey geologists to identify areas threatened by an eruption. Now, other Survey scientists would be watching for signs as to whether the volcano would erupt. To do that, they would have to monitor the volcano's vital signs including its earthquakes, gases, and physical shape.

To monitor eruptions was the reason the Survey's only volcano lab, the Hawaiian Volcano Observatory, existed. HVO was created by geologist Thomas Augustus Jaggar. When he was chairman of the geology department at the Massachusetts Institute of Technology, Jaggar had been one of the first scientists to reach the town of Saint-Pierre on Martinique after the 1902 eruption rolled blistering pyroclastic flows over the Caribbean town, killing all but two of the city's twenty-nine thousand people. Jaggar became convinced that only a permanent volcano observatory could provide the laboratory to understand eruptions well enough to someday make lifesaving predictions. He left MIT and started the Hawaiian Volcano Observatory in 1912 and began recording his observations of the eruptions of Mauna Loa and Kilauea.

"A volcano observatory is always striving to see underground," Jaggar wrote in the journal *The Volcano Letter* in 1925 about the requirements for a productive observatory. "The essential features of a volcano observatory are a volcano, an observer, and the record made by the observer. For success there are needed two things more—enthusiasm in the observer and publication of his record . . . (precise about exact places, maps, times, and measurements). It is this matter of measurement that leads the observer to equip himself with instruments. But the object is not the instrument, it is the measurement. . . . It is not the highest or most famous volcanoes in a given belt that should be selected for steady work, but those that are practicable and reasonably free from danger."

Jaggar found safe and practicable features in Mauna Loa and Kilauea on the Big Island of Hawaii. They erupted more frequently than any other volcanoes in the United States. And they were less dangerous because they were not explosive volcanoes like Pelée but volcanoes that produced thick lava, which geologists call viscous lava, that oozed up through cracks onto the surface and flowed about as fast as a man could walk.

The observatory limped along financially. Jaggar even took to raising pigs to support the facility. But in 1919, it was made a national laboratory.

Over the years, the lab built on Jaggar's genius for gadgetry. His most important contribution, other than establishing HVO itself, was the refinement of the seismograph to record volcanic earthquakes. As magma moves toward the surface, it creates its own earthquakes that are often not felt. By pinpointing the locations of the quakes and mapping the changes in those locations, Jaggar was able to "see" underground in a limited way as magma moved from its reservoir and up through a vent to the surface.

The best of the instruments he helped develop, known as the Hawaiian-type seismograph, was used worldwide by 1928. With this instrument, Jaggar was able to note the arrival times of two of the many seismic waves associated with each earthquake and, based on those measurements, seismologists were able to calculate the distance the quake had traveled to the instrument. With a few of these instruments, seismologists were able to triangulate the distance to the location of the quake.

HVO became the gem of the Survey's volcano program. By the late 1940s, the Hawaiian observatory had become the Survey's primary training ground for its volcanologists, a theoretical lab for exploring new ideas about the mechanics of volcanic eruptions, a place of international collaboration, and a test bed where new monitoring technologies were refined.

Nearly all of the Survey's volcano scientists spent at least a couple of years working around HVO's "red stuff." It was dangerous work. Many HVO men (and they were overwhelmingly men) often pushed the safety guidelines, and they were secretly proud of it.

In the early days, the observatory was always short staffed, so each scientist was forced to learn to use *all* the gadgets being adapted to monitor a volcano's vital signs. Access to constantly erupting volcanoes and to the latest in volcano monitoring equipment meant that work done at HVO was cutting edge. Experience with these new devices produced research papers that regularly found their way into leading geology journals and Survey publications. In the Survey's academic culture, the long lists of published studies by HVO's alumni further enhanced their reputations. This in turn led to promotions so that eventually almost anything having to do with volcanoes was managed by an HVO alumnus. The Survey's view of volcanoes was the view of the HVO grads. They were the tops in their field. When it came to erupting volcanoes, HVO grads felt they had a special, almost intuitive feel for eruptions.

Beyond their rock-solid confidence, HVO scientists shared another characteristic. While field geologists are generally a self-sufficient bunch, often working alone in the wilderness or in a lab, the HVO pros evolved into something of a brotherhood. During Hawaiian eruptions, which were frequent in the 1960s and 1970s, the men worked long hours under hazardous conditions. In that stressful environment, they forged deep friendships. Moreover, in Hawaii, they lived in a remote outpost on the Big Island, so even their families became intertwined. But most of all, they shared a common passion and thrill for their work. They were the best, they were tight, and their bonds remained strong long after leaving Hawaii.

"Feeling earthquakes in the middle of the night," one HVO grad would recall years later, "driving into the observatory, driving out to the Chain of Craters to try to be where the eruption might occur, when lava first broke the surface, it was all a very exciting set of experiences. September seventy-seven, that was my first eruption. It went on for three weeks. Big nighttime fountains. I remember flying over the top of them with Jack Lockwood in his light plane and having [the plane] get shot up in the air by the thermals, and that sort of thing. We were young guys in love with volcanoes and we wanted to see as much red rock as we could. We called ourselves volcano cowboys and we were anxious to go to any volcano that would have us."

If Dave Johnston had gotten a few good years at HVO under his belt, he might have been able to join their ranks someday. But in the eyes of the HVO alumni, Hoblitt and the rest of the Denver group weren't in the running. To many HVO scientists, the Denver group came from the world of old eruptions, cold, dead things, which were about as far from the heart-thumping thrill of red-rock fountains as a geologist could get.

The Denver and HVO groups represented two different cultures.

They even used different tools. The HVO grads were wizards with modern monitoring equipment such as Geodimeters, which were originally designed to measure the speed of light, and tiltmeters, some of which had been adopted from the military, where they were used to level ICBMs in their silos. The boys from Denver, on the other hand, worked mostly with shovels, digging holes.

Moreover, the Denver scientists were unfortunate enough to come from the Survey's Engineering Geology Branch. To most research scientists, engineering is to scientific achievement as wood carving is to chess.

Before March 27, 1980, few HVO grads had even bothered to read the Crandell/Mullineaux paper on St. Helens. For Crandell and Mullineaux to write that this relatively young, steep-sided volcano could soon erupt was stating the obvious. To the HVO pros, the volcano's shape alone (sharp features that time had not eroded) was a neon sign shouting St. Helens was a young, active volcano. The HVO grads didn't devalue Crandell and Mullineaux's work. They ignored it. To the scientists who rushed from one Hawaiian eruption to the next, cold rock records of long-past eruptions paled beside the glow of the red stuff.

The HVO guys shared one other thing, but they rarely talked about it—they were not very good at predicting Hawaiian eruptions. Says HVO alumnus Dan Dzurisin, "We had the pulse of that volcano as well as any in the world, and yet we didn't use the term *prediction*. It was very difficult, almost impossible, it seemed to us, to be specific about when things might intensify and quickly lead to an eruption. It's certainly true that people went home hours before eruptions occurred."

Three HVO grads were among the first scientists to arrive on the scene shortly after the first eruption at St. Helens. Jim Moore and Don Swanson flew up from California and walked into the Forest Service headquarters in Vancouver within hours after the explosion. The two men were at the top of their HVO class. Within days, a third HVO alum, Pete Lipman, would fly in to join them. Each of the three men had a long list of publications to his credit. They were special and they physically separated themselves from the rest to the extent that other Survey scientists working at St. Helens soon began referring to them as the Three Musketeers.

Jim Moore may have been the brightest and most creative of the three. He may also have been the most daring, which was important if your lab is a cauldron of boiling rock. A few years earlier, while investigating oceanic volcanoes, Moore had been trapped on the bottom of the Atlantic. It was 1974 and Moore was exploring the midocean ridge in the deep submersible *Alvin*. The ridge is the place where the North American plate and the European plate are actually being torn apart and the opening wound bleeds lava. Moore's curiosity took him and the *Alvin* pilot nine thousand feet down to where lava was pouring from the earth, creating a bleak landscape of black "pillow lava." *Alvin* held its position on the bottom while Moore took temperature readings. Moore spied a curious-looking formation, and the pilot maneuvered over to it for temperature readings. When the work was finished, the pilot powered up but *Alvin* wouldn't rise. The pilot applied more power, but *Alvin* didn't budge. Unable to surface, they sat on the bottom for hours, trying to figure out what had gone wrong. Meanwhile, the sub's batteries were draining, and without electrical power *Alvin* would remain on the ocean bottom forever. Eventually, Moore and the pilot guessed that they had drifted through this black world and became wedged under something. The pilot tried maneuvering down and sideways, and the submersible broke free. They reached the surface with *Alvin*'s batteries nearly dead.

When Moore first arrived at St. Helens, he wanted to check for signs of ground deformation. In Hawaii, when a plug of magma migrates toward a vent, it deforms the ground ever so slightly. At HVO, scientists had used a Geodimeter to identify and measure these minute changes. But the Survey didn't have a Geodimeter at St. Helens yet. It might take weeks to get one shipped from the Survey's earthquake division in Menlo Park near San Francisco or from Hawaii. Fearing such a delay, Swanson asked to borrow one from the Smithsonian Institution in Washington, D.C. The HVO grad there immediately packed it off to St. Helens.

While waiting for the Geodimeter to arrive, Moore proposed using the snow-covered, frozen Spirit Lake as a big carpenter's level. Moore thought that if magma was moving toward the volcano or if the volcano was swelling from intruding magma, the movement could cause the shoreline nearest the mountain to rise higher than the far shore. Soon Swanson and Moore were trudging around the lake, at times with snow up to their hips, clearing sites, pounding wooden stakes and marking water levels. Days later, they rechecked the stakes and found no change.

That there was no change provided information. It meant that there was not a large, deep body of magma growing or pressurizing under the volcano and driving the steam eruptions at the summit. Instead, whatever was happening was likely happening right under the volcano itself. That's because changes in a large, deep magma body would have caused the ground surface to deform even out to the distance of Spirit Lake (about six miles). Shallow changes would have left the lake unaffected. The "no change" result meant that *if* magma was involved at all, it was already at shallow depth.

They were confident of their results. But, significant changes might already have taken place.

Like Jim Moore, Don Swanson had accepted that working around live volcanoes was dangerous. He had been trapped by a sudden snowstorm near the top of Mount Baker during its emergency. At HVO, Swanson had become an expert in monitoring how volcanoes deform their surroundings as they move toward eruption. The actual swellings and deflations are tiny, but with the Geodimeter, Swanson was able to measure these changes at the surface as plugs of magma migrated across Hawaii's Big Island to the volcano's throat.

The Geodimeter had been a breakthrough tool for volcanologists. The army had used one of the first at their Cold Regions Laboratory at Dartmouth University to measure ice floes creeping in the Arctic. When a Dartmouth geology professor, Robert Decker, heard about the instrument, he thought he could adapt it to detect equally small movements of the ground around a volcano. One summer, when the army didn't have any use for the instrument, Decker convinced the military to lend him their Geodimeter and he took it to HVO. It proved to be perfect for Hawaii because, as was discovered with the new instrument, Hawaiian volcanoes building to an eruption practically hyperventilate.

The Geodimeter works by bouncing a laser beam off a reflector. By measuring the point in the wavelength on which the actual light wave was returned against the hypothetically perfect wavelength, a distance could be measured with phenomenal precision. But the Geodimeter was a twitchy, frustrating instrument that required surgeon's hands to operate and a mathematician's precision to calculate results. The Geodimeter also required two two-man teams to set instruments and record weather conditions (since changes in air temperature along the path could influence the laser beam). Establishing a "line" produced a jumble of numbers from which weather effects and other influences had to be extracted by calculation. In the hands of a Geodimeter wizard like Swanson, the calculations could be done in minutes. But for most others using the machine, reaching a conclusion could take hours. Moreover, since the changes were infinitesimal, each calculation had to be rerun for accuracy.

Shooting a single line was an accomplishment. Detecting a change in that line, which meant reshooting the line, was a test of stamina. But Don Swanson, demonstrating Olympic-class endurance, threw an entire net of Geodimeter lines over Kilauea. His lines were so precise that they became the gold standard for all subsequent distance-measuring devices tested at HVO. But more important, the net allowed Swanson and other scientists to understand the role of ground deformation prior to an eruption.

It's unclear what drove Swanson to do such tedious work, essentially geological bookkeeping on a grand scale, but certainly frustration was a part of it. Swanson had been feeling marooned on that rock in the Pacific while back on the mainland the entire science of geology was being transformed. The transformation was so fundamental and far-reaching that it divided the science into the Old and the New Geology. The two were separated by confirmation of the theory of plate tectonics.

The earth, it has been discovered, is broken into about a dozen fragments, which geologists call plates. By geological standards, these plates are thin, about sixty miles thick. This means we are living on an apple skin crust of earth. Moreover, these plates are floating. In some places, such as the mid-Atlantic ridge where Moore got stuck, the plates are pulling apart, and elsewhere they are banging into one another. All this activity triggers earthquakes and builds mountains.

To Swanson, it seemed that each week the journals would arrive from the mainland with another earthshaking piece of the plate-tectonic theory being confirmed, crushing old ideas and opening up dizzying possibilities. It seemed to settle the long-standing argument about whether the earth was expanding or contracting. It was doing both, in a perfect balance. And it explained why earthquakes occur where they do, along fault lines. Swanson would grab those reports and read them with a sense that he was a bystander during the most exciting geological period he was likely to witness.

When he had arrived in Hawaii, continents were fixed. When he left, it was clear that blocks of the earth were adrift. In every direction Swanson might look from Hawaii to the edges of the Pacific, which had long ago been named the Ring of Fire, plates were busy recycling the earth.

Plate tectonics also pointed to the engine of volcanism. Near St. Helens, for example, about two hundred miles off the coast of Washington State,[2] the Juan de Fuca plate is expanding east and pushing under the more buoyant North American plate as it moves west. As the Juan de Fuca plate slowly dives beneath the North American plate, it carries with it water and minerals from the Pacific's floor. At a depth of 60 to 120 miles, the water, calcium carbonate, carbon dioxide, sodium, potassium, and other deposits act as fluxes that lower the melting temperature of rock. The melted rock increases in volume and becomes less dense. Water and carbon dioxide infuse the liquid rock with gases, making it even more buoyant than the surrounding rock. So new magma begins moving to the surface. It snakes through fractures in the overlying plate. Often these passageways are blocked. Sometimes the magma simply stops rising in these places. But at times, when more magma wells up from below, the magma puts even more pressure on the barrier and shoulders the cold rock out of the way. When magma breaks through the rock like this, it sets off earthquakes or, if it happens at the surface vent, releases an eruption. The debris around the volcanic vent begins to pile up, eventually creating a mountain. This is the process that stoked the fires and built St. Helens and all other subduction volcanoes.

All of the excitement generated by the plate tectonic theory happened while Swanson, acknowledged as one of the field's most promising scientists, was sitting on a rock in Hawaii. In fact Swanson wasn't just sitting on a rock, but at the top of a hot spot. According to the new theory, the Hawaiian islands were created by a mantle plume. A plume is created when a weak spot in a plate allows magma to escape directly to the surface. The rock Swanson was stranded on came from the very heart of the earth. Such discoveries being made right under his feet only made matters worse for Swanson.

The third Musketeer, Pete Lipman, was on a Colorado ski vacation when he saw a newspaper headline about the St. Helens eruption. He quickly got a flight out. Pete Lipman's interest was in the internal dynamics of volcanoes. At HVO, Lipman had become proficient with the three instruments that measured ground deformation: the Geo-dimeter, the spirit level (very much like an engineer's transit), and the tiltmeters. But Lipman could not work long on active volcanoes before becoming bored.

He left HVO to study ancient volcanoes. In 1978, in a place called Questa, New Mexico, he had found a volcano's fossilized innards that had been exposed through a geological accident. Cutaway diagrams of volcanoes illustrate their viscera as a thick, tree-branch piping system leading from a red pool of magma underneath up to the top of the cone. But in a few places on earth, the cutaways actually exist. And not sur-prisingly, they reveal an internal life that is different from what had been imagined. In Questa, near the Colorado border, Lipman had found one of these rare places. Here, 26 million years ago, magma had moved up from its chamber into the vent and then stopped and hardened. That new rock was harder than the surrounding older rock and when a rift cleaved the mountain, it left one side of the volcano's innards exposed above the valley floor. By the time Lipman flew to St. Helens, he was in the early stages of his Questa observations. But it was already obvious to him that magma might indeed move very close to the surface without erupting. He had seen it in Questa.

Almost the minute the three arrived, they decided they couldn't work in Vancouver. Swanson and Moore had actually stumbled into a press conference when they arrived at the Forest Service headquarters. Right then they knew this was not the place they wanted to be. Not only was it an hour's drive from the volcano, it was obvious that a geologist could spend his entire day doing nothing but answering the questions of the uninformed. Too many phones were ringing in Vancouver, and way too many reporters, politicians, sheriffs, and merchants were draining their time. So they drove north, up Interstate 5 until they spotted the biggest motel near St. Helens, the Thunderbird Inn in Kelso, Washing-ton. They chose the motel as the place to set up shop, and they leased a helicopter, flown by a Vietnam vet who used the baseball diamond next to the motel as a landing pad.

Most of the politicians, merchants, and reporters asking questions in Vancouver didn't know that neither the Hawaiian nor the Denver teams knew much about how volcanoes like St. Helens reactivated. Almost no one in the U.S. Geological Survey actually had experience with live, explosive volcanoes. While the Hawaiian volcanoes have exploded violently in the geologic past, all the Hawaiian eruptions monitored by Survey geologists over the previous forty years were relatively benign events. No scientist had seen Kilauea or Mauna Loa blow themselves apart. Instead, the best Survey volcano specialists, the boys from HVO, were trained on and developed instruments for volcanoes that oozed pasty lava, just as dairy farmer and county commissioner Van Youngquist had imagined.

"Very few people there had ever seen an explosive eruption," recalled Thomas Casadevall, who was then a young scientist from HVO. "Many of the people working there had a model of Hawaiian volcanism, which was fluid lava flows that you could walk up to and sample directly."

And that's what made St. Helens such a prize. For the first time, the HVO scientists could use modern techniques and equipment to monitor an explosive volcano as it awoke and, they hoped, erupted. They would measure every possible detail with instruments that were thousands of times more sensitive than the eyes of Pliny the Younger or Frank Perret's teeth.

The HVO scientists had a bedrock belief that a pre-eruption signal existed, and that if they monitored closely enough, they would see it. If they saw that signal early enough, they could do what no one had ever been able to do with one of these explosive giants—they could predict an eruption.

By the end of March, the mountain seemed to be headed full speed toward a cataclysmic eruption. On March 30 alone, St. Helens spouted ninety-three minor eruptions, some with spectacular displays of lightning. One lightning bolt was two miles long. The eruptions triggered rock and snow avalanches, some of which were as wide as a football field and dropped more than a mile.[3] Monitoring small quakes was useless. Malone, the University of Washington seismologist, had no doubt that the quakes were originating right under the northwest flank of the mountain, some of them perhaps as shallow as a half mile down.[4]

The focus of attention, for tourists and scientists alike, was on the eruptions. Some ash pillars rose a mile high. The tallest eruptive plumes of steam, or steam and ash, could reach up to seventeen thousand feet, halfway to a jet's cruising altitude. Some were anvil-shaped and others were blown sharply off the summit by strong winds. The tallest plumes generally consisted of three parts: a lower, very dark finger; an intermediate, gray segment of ash and steam; and an upper cloud of white steam. These steam-driven eruptions ripped rock out of the volcano's throat as they burst out. The rock was heavier than the steam and quickly fell out of the plume and back onto the mountain slopes as ash.

Many of the eruptions popped through the clouds that were often cloaking the summit. When the clouds slid off the summit, airborne observers could see that the top of the mountain was being transformed. The March 27 eruption had blown a hole right through the summit snowcap. On March 30, a second crater was created about thirty feet north of the first. Continuing eruptions widened and deepened the summit crater, and occasionally, snowmelt would create small ponds in the craters. Finally, the wall between the two craters disappeared, leaving a pit six hundred feet wide.

During one of the night observation flights, Dave Johnston saw a leaping blue flame measuring about fifteen feet across. Burning gas had been observed above lava flows and over lava lakes, but never in a crater without lava.[5] The sighting would have helped confirm that this was a magmatic event, but the flame was never seen again.

Dave Johnston's special interest was in volcanic gases. Gas monitoring was a daredevil endeavor even for volcano scientists, because it generally involved going up to or even into a crater to collect samples from fuming vents. Johnston, who had been a runner since high school, was especially suited for sprinting into craters.

The modern era of gas studies had begun with French mineralogist Charles Sainte-Claire Deville. He became interested in the sulfur crystal formations he saw at Soufrière on the island of Guadeloupe in 1842. He continued his research in Italy, working on the 1855 eruption of Vesuvius. Deville repeatedly risked his life to get gas samples. During every phase of the Vesuvius eruption, from its earliest tremors to the climactic eruption and through the decline, Deville kept dashing up the slopes and into the crater of the volcano to retrieve gas samples. Until Deville began his studies, geologists thought that each volcano generated its own unique gas—sulfur dioxide at Etna, hydrochloric acid at Vesuvius, and so on. Deville showed that the gases are the same in every volcano, although their proportions may change from one cone to another.

By the time St. Helens began acting up, scientists were coming to appreciate that volcanoes were really gas-driven machines. When magma is miles beneath the surface, the gases—mostly water, sulfur dioxide (SO_2), and carbon dioxide (CO_2)—are dissolved in liquid rock, like the gas in an unopened bottle of Coke. As the magma rises to more shallow depths, pressure eases and the gases begin to separate from the magma, which is why Coke begins to bubble when the top is snapped. The vapors can separate completely from the magma and leak to the surface, unseen. They can saturate ground water, killing fish in mountain streams or killing surface vegetation including trees. Sometimes, these gases can be released as a large bubble, and because the gases are heavier than air, they can hug the ground and flow downhill, as fog. That happened in Cameroon, West Africa, where a cloud of CO_2 seeped out of a volcano one summer night, drifted along the ground, and before sunrise, suffocated more than five hundred people.[6]

But the most deadly punch from an explosive eruption develops when magma becomes foamy. That can happen when pressure is released suddenly. It's like shaking a bottle of Coke and then snapping the top. Imagine that the foam shooting out at the speed of sound has a temperature of 1,800 degrees Fahrenheit. That's the driver of an explosive eruption.

By the time St. Helens began acting up, another new instrument had become available for studying gases in a less dangerous way than direct sampling in craters.[7] On March 30, 1980, a Dartmouth geology professor, Richard Stoiber, came to Vancouver and brought with him a Canadian instrument known as a Barringer correlation spectrometer, a COSPEC. The instrument, which detected sulfur gas, had been designed to measure air pollution coming out of smokestacks. But again, a creative volcanologist had adapted a new technology to solve a volcano problem.

Low levels of sulfur dioxide (SO_2) are generally found at geothermal sites, but some volcanoes have suddenly released thousands of tons of SO_2 a day, and often those big releases came just prior to an eruption. Generally, SO_2 is dissolved in magma at deep levels, like the CO_2 in a bottle of soda. Very near the surface, large quantities of the gas begin to separate from the liquid rock, which the COSPEC could detect.

Like nearly all the new volcano monitoring devices, the COSPEC was brought first to the Hawaiian Volcano Observatory. At HVO, Tom Casadevall, who was then working part-time for the Survey, became fascinated with the instrument. Stoiber and others had used the COSPEC in spot measurements at threatening volcanoes in Central America. But the scientists would show up with the instrument and take readings for a few days and move on. Casadevall wondered if there were changes over time that might warn of a building eruption. By the time he arrived in Vancouver to work with Johnston, Casadevall had seen that at least some volcanoes could signal their intentions with sudden jumps in SO_2 release.

At St. Helens, the COSPEC did not measure huge levels of gases. Johnston's experiments at the steam vents also failed to show any increase in SO_2 at all. This led Stoiber to conclude that the eruptions, so far at least, were steam-driven and did not involve magma.

The negative SO_2 readings exacerbated the scientists' confusion about what was going on inside the mountain. Perhaps St. Helens was suffering from heartburn. Maybe it was after all just an energetic warming of a volcano like Mount Baker. Stoiber and his students, Stanley Williams and Lawrence Malinconico, made repeated trips to St. Helens and left the COSPEC behind with Johnston and Casadevall. But week after week, neither Johnston nor Casadevall could detect any significant production of SO_2.

As the weeks passed, an organization evolved among the scientists. Johnston and Casadevall would monitor gases. The HVO Musketeers took the bulk of the ground-deformation studies. The seismology was done by Malone and his group at UW in Seattle. It was an awkward arrangement, with groups spread from Vancouver to St. Helens and up to Seattle, which would require close cooperation to work best.

With an organization starting to take shape, Bob Tilling, back at Reston headquarters, called Bob Christiansen, another HVO pro, who had flown to Vancouver from the Survey's Menlo Park, California, office, to ask if he would take charge of the Survey's efforts at St. Helens. Christiansen declined. He would agree to oversee the monitoring work of Moore, Lipman, and Swanson, but he didn't want to be dragged away from the science to do press conferences and advise bureaucrats about roadblocks. He would stay in Vancouver and filter the findings of the monitoring team into the chaos swirling in the Forest Service building.

So, by default again, Mullineaux was left to be the front man. Directing any group of scientists can be like herding cats, but Mullineaux had been put in an especially difficult position. Not only did he have less prestige in the organization, he was a surficial geologist (one dealing with the earth's surfaces, sometimes snidely called a "superficial geologist") from Denver and from the engineering branch. Even his government pay grade was lower than the HVO grads doing the monitoring.

Another problem that was becoming apparent was that people were working themselves to exhaustion because the Survey did not have enough trained manpower.[8] Volcano studies had never been a high priority within the Survey, although its policy of rotating scientists through HVO had paid off with a core of scientists who had some level of monitoring experience. Now, however, the core groups of Survey volcanologists were being stretched to the limit.

Even for those who could wrangle a few days at St. Helens, the Washington volcano was only a temporary diversion. Everyone had other projects with deadlines. Many people stole a week or two from their regular work, and some younger scientists actually paid their own expenses to work on St. Helens. But as time passed without a cataclysmic eruption, they had to return to their real life, too. It was looking as if the St. Helens watch was going to be a long-term affair. Based on the Mount Baker experience, it might be up to a year before St. Helens would settle down or blow up.

Then St. Helens upped the ante. Early one evening, instead of just the occasional earthquake scratching out its transient signal, a strange, continuous, rather monotonous signal was recorded on the seismographs. It started about 7:25 P.M. The trace waved back and forth in a regular way, about one swing per second, and the entire event lasted for five minutes. Dave Harlow, one of the Survey's seismology experts who was visiting UW to help interpret the seismic data, recognized it was a "harmonic tremor." This pattern indicated that molten rock was on the move. About twelve hours later, the same pattern was repeated, this time for fifteen minutes. Maybe St. Helens was no Mount Baker after all. The concentration of earthquake locations under the north side of the mountain meant something was violently pushing aside rocks. No one in the Seattle group doubted that the harmonic tremor was the signature of fresh magma.

New monitoring equipment began arriving by early April. A container of experimental tiltmeters was flown in from HVO along with a team of technicians. The Survey had snatched up the technology as soon as it became declassified. Tiltmeters had been used in the earthquake program for a couple of years, but they were still experimental for Survey volcanologists. The tiltmeters could provide information about ground deformation, rapidly supplementing and extending Geodimeter readings, which could take hours, if not days, depending on the weather. And the tilts gathered information every ten minutes, day and night. But installing them proved to be a challenge.

St. Helens differed from what the HVO alums were accustomed to not only as an explosive volcano, but also as a steep, snow-covered mountain. Survey scientists working in Hawaii had volcanoes with gentle slopes and generally pleasant weather. At St. Helens, they had to learn how to work in harsh weather and on steep terrain. The tiltmeters had to be firmly embedded in rock, which in early April could be under as much as ten feet of snow. Arnold Okamura, who had brought the equipment from HVO, and Don Swanson planted the first one in an outhouse next to the Timberline parking lot. They had to dig through several feet of snow to get to the outhouse door and then chip away at the ice on the bottom. It was a far cry from a nuclear missile silo where they were originally used.

Installing tilt stations almost immediately gave geologists a new appreciation of St. Helens. On April 10, while Swanson and other Survey geologists were working with a spirit-level tilt (like an engineer's transit), the mountain let go with an explosion. As the mountain exhaled a dark cloud of ash, Swanson kept his eyes on the bull's-eye bubble that was used to level the tiltmeter. His experience working with tiltmeters in Hawaii didn't prepare him for what he now saw. During Hawaiian eruptions, changes in tilt were often too small to be detected visually. But as Swanson fixed on the bull's-eye, the damn thing swung away from the volcano and then back. "We watched the bubble in awe for several minutes," Swanson wrote in the monthly report.[9] St. Helens was breathing.

The quakes and steam blasts were happening so frequently that to the scientists working daily on the flanks, these events began to seem inconvenient, no longer dangerous. On April 8, St. Helens pumped out ash and steam for five and a half hours in one sustained eruption. Big earthquakes rocked the mountain so frequently that sometimes they overlapped one another. A plot of earthquakes showed they were concentrated directly under the northern flank of the volcano. Spirit Lake lodge owner Harry Truman moved his bed into the cellar, saying the quakes made him feel as if he were on a ship being tossed in a storm.

The quakes caused large sections of the summit crater's northwest wall to collapse and made the crater grow until it was two thousand feet long and five hundred feet deep.[10] The entire Statue of Liberty and the island upon which she stands could easily have been hidden in the summit crater. Observers in the Forest Service planes could see ponds of muddy water in the crater bottom, ponds large enough to float blocks of ice. With each eruption, the ponds disappeared, exposing a "throat" at least twenty feet wide on the deepest part of the crater floor.

At St. Helens, the main focus of attention—for sightseers and scientists alike—remained the frequent and dramatic eruptions from the summit. Watching those eruptions from her own observation post slightly back from Coldwater was Sue Keiffer. She was a planetary geologist with the Survey, and her husband, Hugh, was an infrared specialist. Both worked out of Flagstaff, Arizona, and they rushed to St. Helens with the report of the first eruption. Their plan was for Sue to carefully document the eruptions while Hugh helped her with logistical support. But shortly after Sue trekked into the snowy ridges around St. Helens, Hugh was drafted into the science group in Vancouver to help establish infrared studies of the volcano. Eventually, Sue ran out of food and had to snowshoe out. But when she returned to Vancouver, she was convinced that—except for the size—the eruptions of Mount St. Helens were exactly like those of the geyser she had been studying at Yellowstone, Old Faithful.

Perhaps the largest eruption ever occurred at what we know today as Yellowstone National Park. Walls of the Yellowstone Valley, separated by fifty miles, were created by that event. The eruptions that produced Yellowstone may have been one hundred times greater than the 1883 eruption of Krakatau.

Like Hawaii, the entire Yellowstone basin overlies a volcanic hot spot. As plates pass over these spots, holes are blowtorched through the plates. The earth beneath Yellowstone is still hot enough to boil water, and that's what happens beneath the park's geysers. Water fills Old Faithful's throat, for example, to a point about fifteen feet below the surface. The water is contained in a long column underground, and the bottom of that column reaches the boiling point long before the top. The bottom becomes heated beyond the boiling point which makes the water column unstable. Then one extra calorie of heat is added or one gas bubble disrupts the system, and it flashes to steam, expanding its volume seventeen hundred times.

From her observation post, Sue Keiffer began to notice that St. Helens erupted at almost regular intervals. She worked out a heat and water budget for the volcano. The process began with glaciers and snow melting as the mountain warmed. Eventually enough water accumulated in cracks, was heated, and boom, another eruption. Each eruption widened the volcano's vent, blowing out more rock, which caused the discoloration of the steam plume, but the widened throat also allowed more ice and snow to fall back into it. While Sue thought all this was interesting, she also thought that the eruptions were almost a diversion. To her, the important action was happening elsewhere.

But summit happenings dominated the daily science meetings in Vancouver. At one morning meeting, after reports had been made on the number and types of eruptions since the last meeting, on the gas and ash content of the eruptions, and on observations made from helicopters about changes inside the craters, Rocky Crandell wrote in his notebook in red pen: "Who is stepping back to take overview?"

In fact, no one was. Crandell was dead-on that details had consumed the group, especially the HVO monitoring team. But this was the way Hawaiian volcanoes were monitored, and those techniques were now being transplanted to the mainland. At the next day's staff meeting, Crandell copied similar observations about the activity at the summit and then wrote: "Activity going on in crater seems to be minutiae, trivia."

Sue Keiffer would have agreed completely with Crandell's observation. Later, when she heard that the volcano pros called the geyser eruptions "throat clearing," she quickly said, "That's wrong." It confirmed the cartoon most people carried in their heads of the innards of a volcano. But the stuff blasting out of the top of St. Helens wasn't coming from a pipe that ran clear to the magma chamber. It was coming from where the rocks got just hot enough to boil water. That hot water, trapped under cold rock and ice, eventually became superheated and found a path out, rapidly expanding, in a scouring eruption, as the HVO pros put it; a geyser to Sue Keiffer.

As the high-pressured days and nights turned into weeks, the scientists grew increasingly tired. Unlike the Forest Service, the Survey didn't have other Rocky Crandells or Don Mullineauxs. Concentration became harder for all the scientists. By now the adrenaline tanks were nearly empty, and the grating differences between the two cultures—hazards and monitoring—were becoming sore spots. Almost from the moment the HVO monitoring pros arrived, they had set out to establish an independent group. They didn't stay in Vancouver, but in Kelso, a small logging town much closer to the mountain. They were asked to report their observations daily to Christiansen, who would feed them into the daily science meeting, but they often failed to find time. Most irritating of all to the scientists in Vancouver was that they had commandeered the only helicopter under contract to the Survey. For the most part, it was unavailable to anyone else. As scientists stationed in Vancouver drove more than an hour each way to change film in an automatic camera on a ridge near St. Helens, sore feelings built up with every mile.

In this stressful time, the cultural differences between the Denver group and the HVO crowd led to disputes. They argued over the HVO group's use of the helicopter. They fought over where the HVO people were living. That irritation grew so serious that Bob Tilling closed the HVO outpost in Kelso and ordered the Musketeers back to Vancouver. This just further exasperated the Musketeers. Minor slights became major insults. In this environment, differences in interpretations were more than just academic disputes. Collegiality evaporated. Malone, for example, repeatedly called Swanson because he wanted to see if his earthquake readings corresponded to anything that the deformation monitoring was seeing; but Swanson never returned his calls.

Some of the difficulties in working together sprang from the character of field geologists. They are independent souls and commonly loners. They are hearty souls who literally live within their work, whether it's under the sea or on a mountain. They have their adventures—of falling through an ice bridge into a deep, jagged crevasse as Crandell had, or being trapped on the bottom of the sea in a small sub like Moore with batteries draining. The field attracts strong personalities and weeds out those who cannot thrive in solitude. While they often feel a sense of collaboration with those who have gone before, such as Pliny, they are unlikely to be good committee members.

Another characteristic of many geologists that made them less than ideal in such a crisis is that they work in different time scales. While most people struggle to grasp the reality of a seventy-year life span, geologists develop an intuitive feel for periods that span millions of years. And they can work in unique spots of time, in some period that produced certain products that were never seen before or after. This ocean of time gives them enormous professional freedom to propose theories. Unlike with surgeons, no patient is at risk on the table. The field allows geologists room to speculate about the origins of the universe from information stored in a single rock.

Normally their ideas and conclusions are presented at scientific meetings where they are attacked by thoughtful critics. So back to the field they go, then back to some other conference with their ideas better refined and supported. And so the process goes, until a paper is published—which can take years. It's generally a small contribution to the field, but it's a link with the long-ago collaborators. Except for rare bursts, the study of geology seems to move only slightly faster than the pace of geology itself.

At St. Helens the scientists were surprised when demands forced them to hurry this thoughtful process. Ideas had to be distilled, debated, and revised not in a few years but in a few days. People who wanted access to their homes or needed to get back to work came up against this closed-door process and beat the door down.

They wanted to know what that mountain would do now. This was something entirely new for the geologists. In the past, they had been called upon to provide the "where" and "what" answers, such as where is the ore deposit and what grade is it. Now, they were being asked to provide the "when," and provide it fast.

As the days turned into weeks, the volcano's eruptions became less frequent, less vigorous. Sue Keiffer speculated that the mountain had about dried out. But the widespread view outside the Survey was that St. Helens was returning to sleep for another century. A *Washington Post* reporter started one of her stories by summing up the popular sentiment: "This apocalypse refuses to cooperate."

It never would, announced a famous French volcano expert. Haroun Tazieff was the Jacques Cousteau of volcanology. Geologists around the world knew of Tazieff. He had been in the business for thirty-two years and had fifteen books to his credit. And he was one of the few volcanologists who had significant experience of explosive volcanoes. Tazieff had correctly predicted that the Guadeloupe volcano would never produce a significant eruption. He came to St. Helens as part of a pre-arranged lecture tour, but when he tried to visit the volcano, he was turned away at a roadblock. Without seeing the mountain close up and without reviewing the Survey's monitoring reports, a miffed Tazieff declared to the press on April 9 that St. Helens would never erupt in a significant way.

Tazieff was critical of nearly every aspect of the Survey's approach. First, he thought that volcanologists and geologists should not be the primary scientists working volcanoes. He had developed a specialized team of volcano watchers composed primarily of chemists and physicists. Tazieff himself was a true pioneer in gas chemistry. He said the Survey's monitoring approach was inadequate in that it didn't incorporate the latest in gas-monitoring technology. He also said that the best indicator of an eruption was a gas temperature of 1,800 degrees Fahrenheit and asked how hot St. Helens gases were. Bob Christiansen reported that the temperatures were much cooler than that but they had not been monitored directly. Finally, Tazieff attacked the importance the Survey was placing on the detection of harmonic tremors, which had appeared to be the one solid indication that the activity was magmatic.

"Harmonic tremors have been closely linked with the rapid transit of molten lava," Tazieff was reported as saying in Sunday newspapers. "But here [in the Cascades] we have extremely stiff magma, which is not at all the same thing."[11]

Tazieff wasn't the only critic, he was just the most respected. General scientists had become critical of the Survey. Their mistaken views, the Survey believed, were formed largely because they were frequently kept away from the mountain, as Tazieff was. But given Tazieff's fame, his criticisms stung. Even so, Survey scientists thought St. Helens had a fifty-fifty chance of producing a magmatic eruption. At a Mexican restaurant, Casa Grande, which had become the Survey's regular dining spot, there was a logbook for guests to sign and make comments. Haroun Tazieff's name (spelled as "Haround" from "Frogtown") appears on April 12 with the comment: "No it won't!" Below Tazieff's name is that of Dave Johnston, who wrote: "Yes it will!"

Actually, even Johnston and his gas-monitoring colleague Tom Casadevall began to think that the St. Helens unrest might not be magmatic. The two men were as expert in volcanic gas readings as anyone in the Survey, and their readings consistently suggested that the St. Helens eruptions did not involve magma.

Not only were magmatic gas measurements a puzzlement, there was no consistent deformation data. Clearly there was some cracking of the glaciers around the summit, but constant snowfalls and repeated layering of ash made any kind of surface measurement unreliable. Yes, there were a lot of earthquakes. The quakes were shallow, but locating the precise depths was impossible. So, the most informed guess at the time led many scientists to conclude St. Helens was just a phreatic, steam-driven event.

At the same time, all the geologists were by now familiar with the Crandell and Mullineaux reports, so they knew that this could be a destructive system if it cut loose. Then people began to question even the historical analysis. What did it mean to have tephra (the stuff that falls out of the air from an eruption) so well mapped? Did a zone of hazard as mapped by Crandell and Mullineaux mean that anyone outside those boundaries was likely to be perfectly safe? Crandell and Mullineaux had mapped the broad range of St. Helens eruptions, from the very large to the nearly insignificant. But no one knew what preceded any of those eruptions. The scientists had no catalog to match these weeks of steam and quakes.

Scientists in meeting after meeting mulled these same questions again and again. New data surfaced and receded, but nothing consistent emerged. St. Helens was a perplexing volcano. None of the models fit. And the longer the emergency dragged on, the more pressure was exerted by local officials and timber industry executives, who all wanted closer access to the mountain.

Working on active volcanoes was a science of recording even the most minute changes and comparing those observations to others taken at other volcanoes. Successfully predicting an eruption would require three things: an understanding of the processes occurring beneath the volcano, monitoring equipment sensitive enough to detect the changes on the surface, and experience of eruptive volcanoes. The Survey had two of three.

The data they had collected by mid-April produced no coherent picture. If magma was involved, it was probably rising, because magma is more buoyant than the surrounding rock. The rising magma should have been releasing its gases, but the monitoring equipment found little evidence of it.

Another way to confirm magma's presence would be if the quakes began to migrate, or move upward. Unfortunately, an earthquake's depth can only be determined reliably when the subsurface environment is well understood. That's because the seismic noise propagates through different layers at different speeds. The stuff under St. Helens was a junk pile of rock slabs, sheets of hardened magma plugs, fracture spaces, and streams. Sound ricocheted every which way. Malone's group was certain the quakes were less than three miles below the surface. But they were guessing when it came to determining how much closer they might be. Given the conditions under St. Helens, the quakes could have been emanating from within the visible cone itself.

As criticisms were lobbed at the historical detective work, Mullineaux questioned why it was taking so long to get anything definitive from the monitoring people. After all, how long does it take to shoot a line or turn an angle, to produce some deformation data? Actually, what the Denver people didn't realize was that the monitoring people were at a loss, too. Accustomed to the Hawaiian model, they didn't know what was important to measure or how often to measure at a Cascade volcano. Moreover, except for Mount Baker, they had never really had to work on a cold, snowy mountain. Ice, for example, complicated the work enormously. You couldn't shoot from one point on the ice to another point because ice constantly moves. You'd get no reliable data. The monitoring group would spend whole frustrating days digging through the ice and snow to plant a rod of rebar in the ground, only to return for science meeting and hear gripes over the delays in getting the data.

Then, in the wake of the Tazieff flap, the director of the Survey made an inspection on April 12.[12] Bill Menard had been a brilliant oceanographer and a teacher who produced inspired students. But at the Survey he was a political appointee, with no experience on volcanoes. The day after he arrived in Vancouver, he was given a briefing, and then geologists piled into a fleet of vehicles that snaked their way up to the Timberline viewpoint so that he could get a close view of the volcano.

Menard rode in the first car, surrounded by the HVO pros. Mullineaux was in the second vehicle. He had been looking forward to the trip, since it got him out of the Forest Service nuthouse and because he had not had a close look at the mountain since the previous summer's field season. The cars pulled into the Timberline parking lot and everyone started to congregate around the director, who was being shown features such as the Goat Rocks dome and the Forsythe Glacier looming overhead.

When Mullineaux got out of his car, he stopped cold. Years of walking on this mountain had made every feature of it as familiar as the furniture in his Denver living room. Probably more so. He saw instantly that the mountain had changed. The monitoring team, all from HVO with virtually no experience on St. Helens prior to 1980, had no basis of comparison. In the last few weeks, Mullineaux had seen pictures taken by observers. But now, standing on the north flank and looking up the mountain, he felt queasy. The upper part of the north face was grotesquely distorted. Above him was a gigantic bulge that looked as if someone had put a fist through the back of the mountain, leaving broken circles of cracked snow and ice above the timberline. This new feature had risen so quickly that it was difficult for a geologist such as Mullineaux to imagine the force that had created it.

"I had been there many times," he recalled, "but from the [aerial] pictures I had just not understood how much change had taken place and how threatening that thing looked. It looked like a failure about to occur. The bulge was terrifying. I wanted to turn around and leave."

CHAPTER 4

The Bulge

Shortly before Menard made his visit, Rocky Crandell invited an expert in landslides from Pennsylvania State University to come out and look at the mountain. Barry Voight looked a little like his brother, the actor Jon Voight, but when he arrived in Vancouver, he was a tad more bedraggled than field geologists generally are. Like Sue Keiffer and the Musketeers, he didn't stay long in the Forest Service headquarters. He rented a car and drove out to the Coldwater outpost where he spent a night with the observer there. The next day, he drove along the ridges near St. Helens until he found a hilltop that allowed him a sweeping view of the north face. He set up a camp, which he called "Coldwater one and a half," and began photographing and sketching the mountain's snow and ice fractures. When he couldn't see the mountain, because of clouds or darkness, he read through the stack of papers on landslides at volcanoes that he had collected before leaving Penn State.

By the time Voight returned to Vancouver, which was the day after Menard left,[1] Mullineaux, Crandell, and other Survey scientists were eager to hear what this outside expert thought about the bulge. Unfortunately, Barry Voight could be tone-deaf in dealing with other scientists. At a noontime science meeting, he offended several of his temporary colleagues when he gave his report.

Voight was convinced that the north slope of St. Helens was primed to give way, or "fail" as geologists put it. The fracture patterns he had sketched and an educated guess about how deep the failing slope would be suggested it was poised to be the biggest landslide in history. What was urgently needed, said Voight, was to get reliable measurements of its rate of movement. If those could be monitored closely, they would warn when the mountain had reached an "inflection point," a sudden increase in the rate the slope was moving. Inflection points were generally, but not always, apparent. Once the inflection point was reached, collapse of the north flank would follow quickly.

Voight also suggested that if there was pressure under the mountain, it could be released by the landslide. The landslide, in other words, could uncork the eruption.

Voight had assessed the HVO monitoring program and figured those boys needed some professional help. They just weren't getting the job done. So, before he made his presentation, he called local surveying companies to find someone who would be available to document the rate of creep. In his presentation, he recommended such help be hired.

His suggestion that commercial surveyors should be tapped for assistance was not graciously received. Could this guy actually be suggesting that the U.S. Geological goddamn *Survey* go out and hire commercial surveyors? Yes, he was, and as a matter of fact he happened to have a list of local contractors who . . . Obviously this guy didn't know the reputations of people like Swanson and Moore. Holy hell, Swanson has laid a net of Geodimeter lines in Hawaii that was so precise it could pick up a friggin' mouse fart.

As the mood of the meeting rapidly soured, no one needed to be reminded that Voight had been brought into their midst by Rocky Crandell. Could anybody take these people seriously? Voight soon left Vancouver to return to teaching classes and writing a report on what he expected to happen at St. Helens.

Even before Voight left, the bulge had become the Survey's main worry. Mullineaux began pressing the monitoring Musketeers for measurements of the bulge. But those were slow in coming. The winter weather now often kept the helicopter on the ground. And when the Musketeers could fly, they found precious little exposed rock. Almost all of St. Helens was under ice and snow. In fact, up through the time of Menard's visit, the HVO team had established only one Geodimeter line. Actually, it was a line that Swanson had done seven years earlier on the volcano's east side, not the north where all the action was. The HVO guys and their field assistants spent days digging holes looking for that seven-year-old benchmark, and when they found it and repeated the measurement, it showed no change.

Three things happened with the recognition of the bulge. First, some of the geologists began going through the literature looking for similar events. One analogue was actually spotted in a 1971 translation of a Russian paper about a volcano on the Kamchatka Peninsula called Bezymianny. Second, starting on April 17, the possibility of a catastrophic failure of the north slope moved to the top of the list as the most urgent hazard. Third, Mullineaux insisted that the monitoring team focus their efforts on the bulge.

Everyone was exhausted. Not enough sleep and too many demands were wearing the scientists down. In this environment, a heated and prolonged argument began over how much risk anyone was allowed to take. This further soured the atmosphere. Worst of all, they could not reach an agreement about what the volcano was likely to do. Then, at one morning meeting, Dave Johnston said he had made an important discovery. He had actually found an eyewitness to one of the mountain's biggest eruptions. He laid his fist on the conference table, opened his hand, and out rolled a toy dinosaur, spitting sparks. Lighten up, folks!

Dave Johnston didn't fit in either the Denver or the HVO camp. He was the youngest member of the Survey at St. Helens, and some people saw the future of the volcano programs in him. He knew explosive volcanoes. He could be paralyzed by public speaking and had a widespread reputation for fainting during his scientific presentations, but Dave would easily tell Hoblitt and the other young Survey scientists interested in explosive volcanoes everything he knew. Over pasta and pitchers of beer, he would sit around at the Spaghetti Factory answering questions and telling stories about his experiences on Alaska's Mount St. Augustine. It was almost a tutorial for those eager to learn about the wildest volcanoes. It was an education for the scientists who would one day take over from Moore and Mullineaux and perhaps combine the two approaches of reading a volcano's history and taking its pulse.

Dave Johnston grew up far from the volcano fields. He was born in Chicago, but by the time he was nine, his family had moved to the blue-collar community of Oak Lawn, Illinois. He had been a Boy Scout who worked part-time as a photographer on the newspaper his mother published and edited. As a child, he dreamed of working for the *National Geographic*. He was also a runner. Lots of things would change in Johnston's life—in college he changed his mind about a future in journalism, he dropped one Ph.D. thesis to pursue another—but he would always be a runner. When he had spent long days working in the field as an assistant, he would still run five miles before going to bed.

While an undergraduate, Johnston worked two summers as a field assistant for Pete Lipman on the ancient volcanoes of southwest Colorado. There was an immediate and strong bond between the two. Dave once told his parents, "I want a job like Pete Lipman's, a home like Pete Lipman's, and a wife like Pete Lipman's." Lipman in turn developed a great affection for his young assistant. The only difficulty Lipman had with the boy was dragging Dave out of the field at the end of the day. Dave was so excited by volcanoes that, if Lipman let him, Dave would work until it was pitch-black.

Johnston pursued his doctorate in geology at the University of Washington in Seattle, but he switched his thesis topic from dead volcanoes to exploding ones after working on the St. Augustine volcano in Alaska. In 1975, Doug Lalla, a friend and former track teammate who had just finished his tour in Vietnam, had moved to Alaska to study geology. He called Dave one spring and said he could get him on as a summer field assistant to work "on a real volcano, not one of those dead ones."

Augustine was a shaking, snorting live volcanic island. On Dave's first visit there, a helicopter dropped the two friends off at the summit and they began their first week of fieldwork. They cooked their dinner in the steam vents. Near the peak, they pitched a tent, which was shredded the first night by a windstorm, leaving them exposed for days to the cold, rain, and fog. Dazed and hypothermic, Lalla fell fifteen hundred feet but stopped himself with an ice ax just before he hit a boulder. Later, Dave and Doug were working at the summit when they were both knocked off their feet by a pre-eruptive earthquake swarm. It was the same phenomenon that St. Helens would exhibit later.

In January the following year, Dave was called back to Augustine. The volcano was exploding periodically and monitoring equipment was in need of repair. A team, including Dave and Doug, took a helicopter out to the island and landed just before an Alaskan whiteout made flying back to the mainland impossible. Trapped in the storm on an island that seemed to be starting another eruption cycle, the helicopter pilot attempted to take off, hoping to get high enough to radio for help. But a gust slapped the helicopter back to the ground—the rotor severed the tail, the skids flew off, and about all that was left was a Plexiglas-and-aluminum ball containing the pilot, who suffered only minor injuries.

The men took refuge in a corrugated-metal shed that had been used as a shelter the previous summer. By now, it was riddled with holes from recent lava bombs, and the front had been caved in by a pyroclastic flow. It was obviously not a safe place to be, nor did it offer much shelter from the blizzard. The stranded men filled the holes, found some canned peaches to eat, and hoped help would arrive ahead of the eruption. They waited for three days. All the while St. Augustine shook and puffed.

Twelve hours after they were rescued, St. Augustine exploded. When Lalla returned to the island, he found the shack still standing, but the mattresses had been incinerated, and the batteries and plastics had been melted. To do that damage, temperatures inside the shack had to have reached eleven hundred degrees Fahrenheit. Nobody would have survived.

Dave and Doug had some excellent adventures on Augustine. Dave burned all the hair off his arm trying to get a gas sampling from a steam vent. As would be the case at St. Helens, he was hoping that the gases would signal that an eruption was on its way. The work he did on those gases, especially chlorine, was some of the best ever done, according to those who picked up his studies in volcanic gases, but he never did detect a signal.

The Augustine experience made him respect the power of a volcano like St. Helens. Some of the HVO pros would later acknowledge that they did not fully appreciate the potential St. Helens posed. "Those of us from HVO really lacked that, we had never really been personally threatened by a volcano as Dave had," recalled Survey geologist Dan Dzurisin. Crandell and Mullineaux may have had a better estimate of the danger, but theirs was abstract, not visceral.

One night toward the end of April, Dave was telling the other young geologists at Casa Grande about the danger explosive volcanoes posed. It had not been a good period for volcanology, he noted. Two scientists, Robin Cooke and Elias Ravin, had been killed in an eruption at Karkar, a volcano on Papua New Guinea, almost exactly one year earlier.[2] And the potential for disaster at St. Helens was high, he said. They were in a dangerous business.

Other people weren't so concerned. By late April, the operator of a Toutle grocery store about forty miles from the peak had sold 360 T-shirts reading SURVIVOR, MOUNT ST. HELENS ERUPTION, 1980.[3] One of them, a yellow T-shirt, became a favorite of Johnston's.[4]

By the third week in April, St. Helens was as much a mystery as it had been almost a month before. The frequency of eruptions had fallen from about one an hour to one a day.[5] The number of earthquakes had dropped to about thirty of magnitude 3.0 or greater a day. The intensity of the quakes, however, had picked up with five to fifteen quakes at 4.0 or higher, so the total seismic energy remained about the same. St. Helens was emitting a modest amount of sulfur gases, but less than 2 percent of what was seen in some volcanoes just prior to eruption. The mountain would tilt this way or that, with no real pattern. Spirit Lake was stable. And the few Geodimeter lines that had been shot showed no change at all. The press was losing interest, and the nation's attention was diverted to Jimmy Carter and Ronald Reagan fighting their way through the presidential primaries.

Right after Mullineaux realized St. Helens had grown a monstrous bulge on its north face, he began demanding that the monitoring team get a fix on it. A week passed and Mullineaux had still not seen any data on the bulge. Then local scientists began telling reporters that it looked to them as if the north side of St. Helens was moving north. This created a stir and added some excitement to the press briefings, but there was little Mullineaux could say. He had no new data on the bulge. In the science meetings, when he asked the HVO pros why there was such a delay, he was told that they were having trouble with the Geodimeter.

Finally, Mullineaux erupted. At one of the daily science meetings, he yelled at Jim Moore, the only HVO pro attending that day, that people's lives depended on the data. Where was it? The day after this incident, another geologist presented Moore and Mullineaux with T-shirts that read EVEN MOUNTAINS CAN BLOW THEIR TOPS. But it would not be laughed off.

"It was terribly frustrating," recalled Tom Casadevall, who was working at St. Helens and would go on to become the acting director of the Survey in the late 1990s. "The weather was terrible. There were very limited viewing opportunities. The environment was covered in snow. And we knew because of Crandell and Mullineaux's studies that the volcano was capable of very explosive activity. We had done these studies of the prehistorical layers of ash, but what did that really mean? You see this layer or this layer or this layer. Is that likely to hurt anybody?

"If there was a feeling of frustration, it was because this was a pretty perplexing system. We were all struggling, trying to figure out what to measure and how frequently do we measure it and what did it really mean? That was the problem. We were coming to a Cascade stratovolcano shrouded in ice with the mental models of Hawaiian-style activity. How do you handle deformation here?

"We were literally staying up after midnight, night after night, talking about what was going on. We simply didn't have the knowledge in the Survey for unraveling this puzzle at that stage in the game."

The HVO pros were struggling with several Geodimeter problems. The instrument is a twitchy device, even in the tropics. In Hawaii, two two-man teams were required to establish a "line." The device didn't give a distance but a list of numbers. When lines were shot, one man would call out readings, which were recorded by an assistant. Measurements of the temperature were taken both at the Geodimeter and at the reflector. The instrument was so sensitive that minor temperature differences between the Geodimeter and the reflector could influence the calculations. Finally, this mass of data had to be reduced and double-checked. It was a headache.

At St. Helens, the scientists faced new kinds of Geodimeter problems. They were working with limited resources. They didn't have whole teams of experienced people to set up, receive, and record, and they didn't have an unlimited supply of expensive Geodimeter reflectors. Also, the Geodimeter couldn't shoot through clouds, which limited the readings to sites low on the volcano. None of the lines showed consistent changes.

In mid-April, it was Jim Moore who was trying to shoot the Geodimeter lines. A brilliant geologist he was, but a Geodimeter magician he wasn't. It took him many hours to work out a few lines, but he didn't admit that to the Vancouver office scientists. By late April, Jim Moore was replaced by Pete Lipman, according to a rotation plan the Musketeers had worked out among themselves.

Moore returned to Palo Alto, where he helped coordinate a photo-mapping project of St. Helens. When the weather finally opened up long enough for mapping flights, the Survey dispatched a plane equipped with large-format cameras to fly predetermined grids over St. Helens. It was precisely the same mission that had been flown the previous August. Those photos would be used to compare the new ones against. The film was rushed to Menlo Park, where each frame was placed in an individual projector suspended above a mapping table. The distance above the table was proportional to the height of the plane. And the two sets of film were projected in two colors. The August 1979 pictures were in red, and April 1980 pictures were in green. On Wednesday, April 23, Jim Moore and his mappers were startled by thick red-green lines that should have been thin and overlapping. The thick regions were places on the north flank that were moved from where they had been just eight months earlier. With a ruler to measure the thickness of the lines, Moore could tell that some points were now 250 feet higher than they were in August.

Something significant had happened to the visible cone of the volcano. Pete Lipman's Geodimeter measurements clarified what that was.

In early April, Lipman had taken two weeks off to return to his office and finish up projects he had been working on. He got back to St. Helens on April 20. The deformation work that had been going on had not yielded anything consistent, but then it was mostly from lines shot low on the volcano. A spot would appear to bulge and then deflate. The mountain may have been deforming like a puffer fish, but it wasn't consistent enough to show any long-term trend that would signal the movement of magma.

With the harsh weather clearing at last, Lipman decided he would try to look higher on the mountain for changes. There may have been substantial changes, but with new ash or snow falling intermittently and clouds frequently covering the summit, such things as new cracks were hidden after a day or two.

Lipman had only a few high-quality Geodimeter reflectors. They were precision mirrors and he didn't want to risk losing them in rockfalls or avalanches. So he bought some thirty-five-cent yellow highway reflectors and attached them to steel fence posts. By use of a helicopter, about a dozen of these targets were planted high on the mountain, some on the bulge and some beside it. Then, from the Timberline parking lot, Lipman shot Geodimeter benchmark lines to the reflectors.

Based on his Hawaii experience, he expected that it would take at least a month to get a meaningful change. At HVO, scientists thought deformation was fast if Geodimeters caught changes of an eighth of an inch per day.[6]

Four days after Lipman shot his first lines, he decided to check his measurements. He drove up to the Timberline parking lot with a field assistant and set up the instrument. He entered the coordinates of the first target into a spotting telescope and looked for the reflectors. No target. Like Moore, Lipman had not shot a Geodimeter line in a few years and wasn't sure he had remembered the process correctly. He moved the telescope a twitch. No target. "Oh my God, I've completely blown this," he thought. He moved the spotting telescope some more. No target. He began thinking that maybe the target had been taken out by an avalanche. So he began scanning the area, and then he saw it. It was a full diameter outside the viewing scope.

It was the same with the second target he tried to find. The consistency indicated he may have made a systematic surveying error. Not a pleasant thought for a Survey geologist. Maybe Barry Voight was right. Then he set the coordinates for a third target. It was right where it should be. He shot another target. It, too, was right where it was supposed to be. A fantastic idea began to form in Lipman's mind. He shot another line, and another. The conclusion was staring him in the face. Four sites clustered in an area about a mile wide on the bulge had moved; three sites outside that area had not. A large but discrete portion of the north face had moved, perhaps as much as several inches, in just a few days. That explained why there had not been any consistent changes recorded by tiltmeters or even the Geodimeter pointed lower on the volcano. The action was high up on the volcano, on the bulge, and it was limited to that area alone.

The questions began bursting in his mind. He started twisting dials, calling numbers, and working calculations. How big was the area that was moving? The Geodimeter lines indicated that stations on Goat Rocks, The Boot, and North Point were all moving. But targets to the east and west of those points remained stable. So the bulge on the move was about a mile wide and a mile long. The sharp boundaries around the bulge indicated that whatever was pushing the bulge was not far beneath the surface.[7]

How fast? Lipman and his assistant finished their rough calculations and found that, give or take an inch, in the last four days a substantial portion of the mountain had moved twenty feet. That couldn't be right. He reworked the calculations and got twenty feet again.

What direction? Ground deformation around Hawaiian volcanoes is a lot like the chest of a sleeping man—a slow rising and falling. St. Helens, he realized, was not acting like a Hawaiian volcano. It didn't deform either up or down. One entire sector of the mountain was heading north.

He jumped into his truck and raced back to Vancouver, ran into the Forest Service building, and told the first group of geologists he spotted, "Guys, we've got us a real problem here."

"We were shocked," said Lipman years later. "We got changes like that in Hawaii in connection with a magnitude 7.2 earthquake, but never in connection with a volcanic eruption. This was just totally outside our experience."

The next day, Lipman found the mountain had moved another five feet. And it moved five feet the next day. It wasn't moving up. The bulge was moving north—sideways—toward the Timberline parking lot, toward Spirit Lake, toward Coldwater Ridge, toward Mount Rainier, at about five feet a day. The moving bulge was like a swimmer leaning farther and farther over the edge of a diving board.

"What it was shouting at us," said Lipman, "was that the bulge was exceedingly dangerous, and unless it stopped, it was going to fail."

It didn't take Crandell long to estimate the size of the landslide that could let go and what that landslide would do downslope. When it did, Crandell said, it would carry with it glaciers, forests, and boulders. Driven by gravity, it would sweep over Timberline parking lot—and the scientists working there—in a few seconds. The landslide would continue right over Harry Truman's lodge and plunge into Spirit Lake, perhaps even riding up and over the hills on the lake's northern shore.

On April 30, the public was alerted to the danger. The Survey announced that an area on the north slope had moved more than 320 feet since last August and the danger of landslide "now represents the most serious potential hazard" posed by the volcano. Rocky Crandell warned that tons of debris could drop from the 7,600-foot level to Spirit Lake and generate a large wave. "Such an avalanche could move downslope at a high velocity—more than one hundred miles per hour—and could move long distances," said Crandell.[8]

View from timberline: The bulge developing on the north side of Mount St. Helens as magma pushes up within the peak. (PHOTO: PETER W. LIPMAN)

The warning did not alert the public to the worst possible danger from a landslide—that the avalanche could uncork an eruption. But many scientists at Vancouver believed that to be a possibility.

For safety reasons the Denver team now wanted the Timberline parking lot to be abandoned by the monitoring team and observations posts moved farther back. This led to another argument between the two teams.

The HVO grads had been working for weeks under dangerous conditions. It was no picnic. On good days, the fallout from the constant phreatic eruptions rained mud on them and their instruments. Working there was like trying to do science while standing on a trampoline. They would feel the earthquakes rolling in. The big ones, they called "leg spreaders," because that is how they tried to keep their $25,000 Geodimeter from crashing to the ground. The big quakes in bad weather were the worst. At those times, the scientists standing in the Timberline parking lot usually heard avalanches roaring down the mountainside.

There was nothing to do but stand there and hope it would stop before it buried them.[9]

The HVO people knew it was dangerous to monitor volcanoes. Hadn't they faced those threats in Hawaii? This mountain may have been dangerous, but they also thought it was a scientific treasure.

The HVO grads figured the Denver hazard scientists simply didn't understand that of course the job was inherently dangerous. But the HVO grads had never had an opportunity like this and it might never happen again. They had to be on the mountain. For the Denver scientists, everything had been learned that could be learned by looking at the geologic record.

"The thing we couldn't believe," recalled Dan Miller, "was that [the HVO guys] didn't understand what an explosive volcano was all about. They were running all over the place, basically standing in the gun barrel, looking up into the cylinders and saying, 'Wow, isn't this great!' "

The hazard and the monitoring crowds had struggled in a forced marriage ever since the Musketeers were ordered back to Vancouver from their Kelso outpost. The main reason for that move, it appeared to the Musketeers, was to have them attend the tedious daily science meetings. This new demand to abandon their viewpoint at the Timberline parking lot seemed to them to be little more than a power play driven by people who were uniformed about "real" risks, which the Hawaii-trained scientists knew well. To the HVO guys, who were working under the bulge every day, the hazards group were just office scientists who spent their days talking to reporters and bureaucrats.

Rocky Crandell didn't see it that way at all. The danger presented by the bulge to people monitoring at Timberline was, to him, exactly the same situation he had struggled with time and again in Europe during World War II. Even years later, Crandell would speak of the monitoring team at Timberline as the "forward observers." He felt the responsibility of an officer, which he was during the war, was to protect the lives of the people ordered into dangerous duty.

Crandell never talked about forward observers, and the HVO scientists never called the demand a bald power play. So they argued about other things, including what the volcano's symptoms meant. And the monitoring at Timberline continued.

One of the arguments centered on what was driving the bulge. In this as in other things, there were several camps. Mullineaux thought that the bulge was created back on March 27, during the first eruption. What the instruments were measuring now, he argued, was a slow landslide. And that land mass would reach a failure point and collapse in a monster slide.

Lipman disagreed. Back at his dead volcanoes in New Mexico, Lipman had seen an internal process he didn't understand then. Now, seeing the process develop before him, it seemed clear. Volcanoes are not straight pipes to an underground reservoir of liquid rock. Explosive volcanoes such as St. Helens and the Questa cones were junk piles. As magma rises, it is fighting against cold, hard rock overhead. Essentially, the weight of the entire mountain is pressing down on it. Sometimes, a mountain of overlying rock just refuses to be pushed aside. And as magma expends energy in this fight, it loses some of its gas, heat, and buoyancy. Magma can reach a point where it has expended so much of its energy that it cannot overcome the pressure of the overhanging rock.

The stationary magma can begin to cool in place. As that happens, the magma forms a dome hidden *inside* the mountain. This is called a cryptodome. Lipman had seen just that at Questa, and he argued that was what was happening inside this volcano.

Two things can happen at this point. The cryptodome can grow thick and solid, essentially permanently capping the potential eruption.

Or fresh magma can continue to rise and exert increasing pressure. This increases the pressure of the trapped magma and gases, and it can cause the mountain to inflate like a balloon, producing a bulge. If the rocks above the cryptodome slid away, the pressure holding back the rising magma would be released exactly like a cork from a bottle of champagne.

Another possibility came from Crandell, who thought a slide could happen independent of an eruption, neither caused by nor causing an eruption. It was a hazard all by itself.

And yet another possibility was considered by a few, including Dave Johnston. They suggested that the landslide could uncork an eruption, with the explosion pointed *out* from the new hole, as if a cork were popped out of a bottle that had been lying on its side. Those did occur. They were known as directed blasts. But they were extremely rare. A Soviet volcanologist had reported on one that was thought to have occurred at Lamington, in Papua New Guinea, killing three thousand people in 1951. A second possible example reported by the same Soviet scientist occurred at Bezymianny, a volcano located on the Kamchatka Peninsula, in 1956. But the author of those reports had not directly witnessed either event.

All possible courses were proposed and debated. Evidence for one or another was put forward and then challenged. It was the kind of intellectual mud wrestling that is the soul of science. But while the process is generally done in the pages of journals over years, these ideas were mulled in one small conference room over a few weeks.

The public however was never allowed to know the depths of the division in the science group. The geologists didn't want to ignite the kind of public debate that made Guadeloupe such a disaster. It was bad enough that local geologists were passing themselves off as volcano experts and issuing their own warnings. The Survey was not about to elevate the legitimacy of that wrangling.

So at the daily press conferences, the geologists focused on the "most likely scenario." And the most likely scenario at the end of April and in early May was a landslide that could cause a tidal wave in Spirit Lake, possibly followed by a vertical eruption with ash and mudflows that would be channeled down the surrounding valleys.

While the Survey geologists argued among themselves about the less likely scenarios, they all believed that before the action started there would be some obvious change on some monitoring instrument, a signal Barry Voight had called an inflection point. That signal could provide some warning time, hours or just minutes.

Inflection points were common features of landslides. Land masses are not perfectly at rest one second and falling the next. Slides begin with small movements, often so small they can't be observed without instruments. But with careful monitoring, landslide experts told the Survey scientists that just prior to the big failure, the rate of slippage should accelerate. The point of acceleration was the inflection point, the geological fire bell.

The geologists became convinced they would see an inflection point prior to an eruption. American scientists had never wired an explosive volcano with more modern equipment than they had at St. Helens. The volcano was covered with seismic nets and, at last, by a Geodimeter's distance-measurement lines. Tiltmeters measured the volcano's hiccups and COSPEC measured her sulfur exhalations. U2s had flown infrared imaging missions, and who knew what the spy satellites were picking up.

One of these instruments would record the signal. It was that belief that kept the students working around the clock, week after week, at the University of Washington's seismology lab. It was that belief that drove the HVO pros to risk their lives every day at Timberline. The signal that observers had looked for since Pliny the Younger would be there, in some squiggle on one strip of seismograph paper or buried in a dense set of Geodimeter calculations. It would certainly be there.

While many of the scientists were poised over their instruments, scanning for the telltale blip, Rocky Crandell had another job. He had to convince officials of the hazards that they were facing. If his science was to be useful, it had to be understandable to decision makers and it had to be compelling enough so that they could make wise decisions even in the grips of political and economic pressures. Science, Crandell thought, doesn't exist for scientists alone. Unlocking the mysteries of nature was an intellectual quest, but providing expert advice to those at risk was, to Crandell's way of thinking, the work's reward. To have value, science had to be converted into action.

The law enforcement officers and officials with the state's Department of Emergency Services were already believers. But sightseers, the timber companies, and the Spirit Lake homeowners were pushing closer into the restricted "red zones." Convincing people that St. Helens was dangerous became increasingly difficult.

By April 20, the official start of the fishing season, the mountain had stopped spouting a steady stream of eruptive columns. Now that the weather was warming, tourists were arriving by the busload to see the volcano pop, but they were often leaving with a picture of an eruption on a St. Helens T-shirt, not in their cameras. Sue Keiffer thought the steam eruptions were stopping because the volcano had boiled off all of its available water. St. Helens had dried out. It was still just as dangerous. Probably more so. But with the eruptions tapering off, the perception grew that the mountain was becoming less risky.

To local loggers and cabin owners, there was another sign that the mountain might not be as dangerous as the scientists would have the public believe. It was Harry Truman. On May 1, when the governor signed an order establishing restriction zones around the mountain, the state attorney general said the order would probably not apply to Truman.[10] The rumbling mountain was turning a crotchety old recluse into a folk hero. To many of the geologists who had worked around St. Helens for years, Truman was a hot-tempered, foulmouthed drunk. Harry said he "led a life as clean as angel's drawers." Harry had his charming side, as some young female Forest Service employees knew. Mostly, though, Harry was a man in grief, still stinging from the sudden death of his wife and comforted, perhaps too long, by his twenty-eight cats and an endless supply of whiskey.

Reporters were delighted with Truman. The old man was an endless storyteller and some of his stories could have been true. He said he staked out his forty acres on Spirit Lake in 1929, after he quit running whiskey from Canada to California. He started his operation by renting tents and guiding campers into the region, and that business had grown into a full resort with a massive three-story log lodge, lakeside cabins, and a small fleet of boats and canoes. Over that time, Truman had survived two-hundred-mile-an-hour windstorms, a fire that had destroyed his first home, and two earthquakes that had toppled his chimneys. He said that in case of real trouble, he had a secret hideout, an old mine shaft he had found years ago deep in the hills.

"I don't have any idea whether it will blow," he pontificated one day in his kitchen while holding a Coke glass probably filled with whiskey. "But I don't believe it to the point that I'm going to pack up."[11]

To most people who lived around St. Helens, it seemed impossible that Harry would have been allowed to remain at his lodge if there was a real danger.

Local sheriffs knew that trying to forcibly remove Harry from his lodge could lead to violence. Moreover, they were not sure of their right to keep anyone from a volcanic hazard zone. A high-visibility confrontation could have exposed their weak legal authority and made for larger problems with the general public. And even if they could force Harry to leave without shots being fired, there was really no way of keeping him from returning. So Harry stayed, gave lots of interviews, collected his newspaper clippings, and created the impression that the volcano was not dangerous to those who had the gumption to stand up to both the mountain and weaselly federal boys.

And there was another sign that officials didn't consider the danger the mountain posed as life-threatening. Loggers were allowed to cut trees right on the slopes of St. Helens.

Logging on Forest Service land had been stopped in March. But much of the land on and around St. Helens was owned by private logging companies. Governor Dixy Lee Ray and Forest Service supervisor Robert Tokarczyk established red and blue zones around the mountain. The red zones were off-limits to all but scientists. The blue zones could be worked by loggers, and property owners could enter with special permits, but no overnight stays were permitted.[12]

"We're logging ten miles away from the peak," one logger told a reporter on May 4, exactly two weeks before the catastrophic eruption. "I don't see any hazard."[13]

And finally, the Survey itself was not being entirely forthcoming with the public. They consistently presented the group's "most likely scenario." The most likely scenario was the consensus of the group, if a consensus could be reached. If not, then the most likely scenario the public was told about was the speculation that fell midway between the extreme possibilities. But the entire range of possibilities and the relative probabilities were rarely discussed. Dissenting views and conflicting evidence were not emphasized in public.

When there is disagreement behind the scenes, most scientists tend to hide it from the public. They do this for three reasons. First, they are certain they know what is best for the rest of us or at least what is better. Second, they believe their science is too complex for the public to understand well enough to make well-reasoned decisions. Third, scientists rarely trust journalists. The two groups, while both relying on communication for their work, distrust each other's core beliefs.

Scientists carefully question, objectively analyze, and release their conclusions into the scientific community to be scrutinized, criticized, and either validated or discarded. Scientists consider it an almost sacred process, and it has worked wonders. Journalists, however, have faith in another process, often expressed as, "In the marketplace of ideas, the truth will survive." This is a much messier process, which occasionally leads to decisions that are harmful to society. The process also works to expose bad decisions, so in the end, it, too, has worked wonders for society.

Scientists often look in horror at the marketplace of ideas. Journalists, for their part, see these scientists as paternalistic elitists who lack an essential faith in democracy.

It's a predicament that scientists frequently confront, and having more faith in science than democracy, they tend to fall back on the culture they know best. Not given sufficient information in official briefings, reporters on the St. Helens story could not know what the possibilities were. And it was not often productive to go to outside sources. Although a lot of outside geologists lined up as volcano experts, there was even less expertise on explosive volcanoes from American scientists outside the Survey. The result was that few people—those living near St. Helens, sightseers, and even some public officials—were fully informed of the possible courses St. Helens could take.

All of this—the limited public disclosure, the logging locations, Harry Truman's defiance, and the quieting mountain—added up to make the Spirit Lake cabin owners increasingly angry about the police roadblock between them and their houses. When their property-tax bills started showing up in their mailboxes, the owners felt that now they even had to pay for the right of being denied access to their own houses. They had all heard about the possibility of a landslide, but their homes were five miles or more from the summit. Snow avalanches rarely reached Timberline parking lot, much less another four more miles downslope.

Many of the lake places were vacation cabins, used primarily during the fishing and hunting seasons. But nearly all the cabins had something of value—a snowmobile or granddad's double-barrel or a favorite quilt—that the owners couldn't retrieve. So the cabin owners protested at roadblocks and at every public meeting held to discuss the state of the volcano. Lawyers called the fire boss, Paul Stenkamp, and demanded to know what legal authority he had for keeping out the homeowners. "My stock answer," said Stenkamp, "was 'Go find out what I'm doing that's illegal and call me back.' "

Determined sightseers, who didn't have to haul out bulky belongings, didn't even bother to fight the authorities. They began skirting the roadblocks by taking any of the logging roads that laced the region like a spiderweb. Soon, geologists working high on the volcano were meeting up with backpackers descending from the summit.

"No one would listen," Skamania County sheriff Bill Closner later told a Vancouver newspaper reporter. "It didn't matter what we did.

People were going around, through, and over the barricades. Some were climbing right up to the rim of the crater."

Even though the eruptions were slowing, the volcano was showing new signs of unrest. On May 6, planes from the Navy's 128th Attack Squadron based on Whidbey Island flew an infrared study for the Survey and detected a small but previously unknown area of warm rock on the north slope. The warm spot was about one hundred feet long and fifty feet wide. It was right in the middle of the bulge. Steam eruptions resumed on May 7, and two harmonic tremors were recorded the following day. Given these signs, geologists began to feel more confident that they would be able to see the inflection point.

The Sunday newspapers of May 11 reported that Survey scientists said they might be able to give an adequate warning of a major eruption. They also announced that at the first of the month, they had established a new observation post, called Coldwater II, on a ridge 5.7 miles north of the summit.[14]

Coldwater II was two miles closer than the first Coldwater observation point, which had left Hoblitt shivering and shaking in the early days. With a new repeater installed, it had a better radio link to Vancouver. Coldwater II would be manned twenty-four hours a day, and observers would be able to issue early warnings of landslides, mudflows, and pyroclastic flows threatening residents downstream on the Toutle River. Even though Coldwater II faced the bulge almost head-on, Rocky Crandell and Dan Miller had done a special analysis of the volcanic deposits on the site and found that, except for ash falls, the ridge was high enough to have escaped every eruption in the last thirty-eight thousand years.

Hoblitt's hand-me-down trailer had been hauled up to Coldwater II. The observation post was staffed most of the time by Harry Glicken, a wiry young geology graduate student who had been hired by Dave Johnston to be his field assistant. Harry was disorganized, often disheveled, and so inattentive to the normal minutiae of everyday life that he was a menace behind the wheel. But he was enthusiastic, a careful observer, and a promising scientist.

Miller and Hoblitt tried to replace Glicken with a video system. It was Hoblitt's idea. He had made inquiries and found a system that would work, with a microwave relay to Vancouver, and a company that would install it. The company set up a test, and the video system demonstrated itself to be a pretty good observer. The price, however, was around $40,000. Hoblitt asked for the money and was turned down.

The St. Helens operation was about to break the bank. Helicopter time alone was becoming more than the Survey could manage. Bob Tilling's personal office account was now over a million dollars. There was a real possibility that Tilling could be charged with grossly mismanaging that account, and at a minimum the charge could cost him his career. Now Hoblitt was asking for $40,000 for a television system! Now, of all times, just when it seemed the mountain was dying down and probably never going to erupt. The Survey was spending $2,000 a day to support the work at St. Helens, which didn't include the $300 an hour for helicopter support, and warning, "We may have to cut down."[15] (The Forest Service, in contrast, was spending $2,000 a day just on aerial observations and was not preparing cutbacks.) In that context, the request for a television system was denied.

Hoblitt and Miller looked for an alternative. Soon, Miller managed to talk the U.S. Army Reserve into lending him an armored personnel carrier to be placed next to Hoblitt's trailer. Miller reasoned that in it Glicken would be protected from any flying debris kicked up by an avalanche or an eruption, and a little flying debris was probably the most hazardous thing Glicken would have to face on the Coldwater ridge.

St. Helens had her vital signs as well monitored as those of a patient in an emergency room. Human observations were being made constantly, not only from Coldwater II on the north side but by the Forest Service, which was flying nearly around the clock, as well. A high-tech network of Survey seismometers, tiltmeters, gas-measuring instruments, and other monitoring equipment spread a net over the mountain to detect any precursory quaking or slipping that could trigger an eruption.

"There is a good chance we would have warning before a serious eruption," Mullineaux was quoted as saying on May 11.[16]

The scientists' only caveat was that if the changes occurred while the mountain was covered in clouds, warning might be somewhat delayed.[17]

With newspapers reporting that level of confidence and the growing pressure from the public, in the second week of May, state officials agreed to allow greater access to the volcano. On Friday, May 16, workers were allowed to remove equipment from the Boy Scout and YMCA camps on Spirit Lake. This action angered the Spirit Lake cabin owners, who threatened to go through the roadblocks "come hell or high water."[18] The homeowners turned to the Survey to ask if it would be safe, but Dan Miller told them no. Given the continuing growth of the bulge, he said, it was "too great a risk." Miller felt that the pregnant mountain looming over Spirit Lake might let go without any warning.[19] Because of the threat of violence, however, the homeowners were given permission. The state police announced that on the coming Saturday and Sunday, they would escort a few dozen cabin owners down the narrow Toutle River valley to Spirit Lake to retrieve their belongings, and escort them back out.

With the scientists growing weary, the Survey decided to begin a rotation of its senior people by the second week of May. Pete Lipman returned home, leaving Don Swanson behind to keep the monitoring network humming. Bob Decker, the scientist in charge at HVO, returned to Hawaii. Rocky Crandell sent Rick Hoblitt back to Denver to write a paper on one of St. Helens's most recent eruptions, and then Crandell went back to Denver himself. Don Mullineaux flew to southern California to attend his daughter's college graduation. On the weekend of May 17, the only member of the Denver hazard team to stay behind was Dan Miller.

At the University of Washington, however, there was no pulling back in the seismology lab. Scientists there saw the eruption in purely seismic terms, as blips on their recorders, and they saw St. Helens as a very active mountain. Yes, the number of quakes had dropped dramatically, but the quakes that continued were more intense. So the total level of energy rumbling beneath the mountain remained constant. No inflection point was seen in the seismic readings, but certainly there was no decline of energy either.

With most of the Survey leaders out of town, the tension in the daily meetings was considerably eased. True, the remaining scientists could read the data as well as anyone, and to be sure the situation was growing more dangerous every day. But the quieting volcano began to deceive many of the Survey geologists.

On Thursday night, May 15, Sheriff Les Nelson drove out to Spirit Lake to try one last time to talk Harry Truman out of staying. Nelson was worried about driving the narrow valleys with a mile-wide bulge hanging over the road and Harry's lodge. St. Helens now frightened the county sheriff. When he arrived, he positioned his car pointing down the highway and left the motor running. Truman told Nelson he was wasting his time. There was no way Harry Truman would leave this mountain. Nelson felt he had done his duty and got back in his car and drove away fast.

After midnight, a large twin-engine plane contracted by the Department of Energy flew out of Las Vegas and headed to Mount St. Helens. The plane and the instruments it carried were part of the black world of the Energy Department. It carried advanced infrared cameras that could detect heat from a cup of coffee from five thousand feet away.

In the hours just before sunrise, Friday, when the ground was as cool as it was going to get, the pilot triggered the camera. Strip after strip of data were made as the plane made nine passes over the mountain.

Before noon Friday, the flight had returned to Las Vegas and the Survey was asked what should be done with the infrared film. It could be processed over the weekend, but that meant bringing in a team to do it, and that would be expensive. No, don't do that, the Survey scientist replied. He was concerned about cost, and there appeared to be no urgency. It could wait until Monday.

The camera recorded an unusual and significant increase in heat coming from the mountain. Measurements of the heat put the energy at two megawatts, enough to power a thousand homes. This amount of heat is rarely observed in nature. An infrared expert with the team speculated later that the heat was either from rising magma or hot gases or both.

Also on that unprocessed film was what an internal report written later called "a strong new feature." The heat was coming from a number of new hot spots located along the upper edge of a bulge. According to the report, the infrared images showed that the new hot spots dotted the upper portion of the bulge as if the mountain had been "perforated." Something profound was happening on the high border of the bulge.

Also on that Friday, Barry Voight's final report had at last arrived in Vancouver. It had taken Voight longer to finish than he had planned. First, he struggled with the calculations, and then there had been a bottleneck in the Penn State typing pool.

Voight, the landslide expert, described a potential collapse of the north face of the mountain in a massive avalanche that could trigger a catastrophic blast from the volcano—precisely what was about to happen.

Crandell wouldn't see Voight's report until he returned to Vancouver, on May 18, after the massive eruption had started. Others in Vancouver did see copies of it before May 18, and it was circulated to the Forest Service and the governor. But it was only one more guess about the course St. Helens could take. The volcano experts in Vancouver certainly did not consider it the most likely scenario.

On Saturday, May 17, 1980, the day before the catastrophic eruption of St. Helens, Rocky Crandell sat down in his home on the outskirts of Denver and began typing a family letter about St. Helens.

"I learned last night," Crandell wrote, "that eruptive activity has again temporarily quieted down though the earthquakes continue at the same rate and the volcano continues to 'bulge.' There is little question that molten rock has moved into the volcano, but there is still some uncertainty about when (or if) it will come to the surface.

"Our big concern right now," he continued, "is whether an initial eruption of molten rock will be relatively quiet and form a dome or lava flow, or will be explosive and produce a big volume of pumice which could then be carried by winds for hundreds of miles from the volcano. St. Helens has done both of these things repeatedly in just the last five hundred years. And it is knowledge like this—to know what happened in the past—that has been the goal of our studies of Cascade volcanoes over the last thirteen years.

"My private concern, which I don't intend to include in a press release, is that the continuing strong earthquake activity indicates that a large volume of molten rock is continuing to move into a reservoir beneath the volcano. The more molten rock is in the reservoir, the larger will be the eventual eruption."

In the six weeks since the first earthquake, much had been learned about the state of the volcano. The Survey scientists were confident that magma was involved. They were also confident that magma had moved close to the surface. They had identified the new bulge and clocked its speed of growth. They also knew that the bulge could not continue growing indefinitely.

But the day before its cataclysmic eruption, no Survey scientist had an inkling that St. Helens was so close to erupting. They were debating whether the biggest threat was from an eruption or from a landslide. They were beginning to believe the eruption, whenever it happened, would be big, although there was no consensus as to how big. The scientists didn't know if the coming eruption would be violent or quiet.

Just before the eruption, little of all that had been learned was helping to warn those living and working around the volcano that St. Helens had become a highly pressurized container made of rock. And the rock container was cracking. For the scientists involved, the day before the eruption, May 17, was in fact like any other day.

Field Notes

MAY 17, 1980

MINDY BRUGMAN

Saturday morning in Vancouver started with a meeting of the remaining volcano scientists in the Forest Service building. The first order of business was to replace Harry Glicken at Coldwater II. Glicken had to go to the University of California, Santa Barbara, for a few days to discuss the graduate work he hoped to begin there in the fall.[1] Observations from Coldwater II had to continue, so a replacement was necessary. Don Swanson agreed to replace Glicken for the few days he was to be gone.

Measurements of the volcano's vital signs were reported. The previous day's reading from the Coldwater II monitoring instruments showed the bulge expanding at a rate of five feet a day, but no measurement showed a significant increase in that rate. If anything, there might have been some slowing. No big changes in tilt or seismic activity. Dave Johnston noted that there continued to be an increase in sulfur gases, but nothing dramatic. No new infrared data were available to report. In sum, all monitoring parameters were essentially the same. There was no inflection signal anywhere.

Before going to the mountain to begin monitoring, Don Swanson stopped Dave Johnston in the hallway and asked if Dave would replace him for one night, that night. Swanson had a graduate student with him from Germany. The student would be leaving Sunday morning, and

Swanson wanted to make sure his student took off okay. If Johnston would take over Coldwater II for Saturday night, Swanson said he would relieve him early Sunday afternoon.

Johnston was reluctant. Unlike Swanson, he believed St. Helens to be extremely dangerous. Johnston would risk his neck when he thought he could get good data. In fact, today, he was planning to go into the crater to collect some gas samples. But just sitting on a ridge, with the bulge practically pushing directly at you, no, that wasn't anything he really wanted to do. Swanson persisted, and Dave finally relented. Perhaps he felt an obligation since Glicken was his field assistant and therefore his responsibility. So he said yes. Before leaving, he called his girlfriend in Palo Alto and told her about the change in plans and how exhausted he was. "God," he said, "I don't know if I can even drive out there."

And so began another ordinary day around St. Helens.

About 10 A.M., a caravan of fifty cars loaded with homeowners was escorted by state police to their cabin around Spirit Lake. They had won their bid to return to retrieve belongings. The agreement allowed them access for a few hours today, Saturday, and again on Sunday morning.

One of the cabin owners, who was loading a Foosball table and lawn chairs into her pickup, was asked by a reporter if she was ready to sell her spot on the lake. The home, which had a picture-perfect view of St. Helens framed through the tall fir trees, was miles from the mountain. The woman said, "I still can't believe in my mind that anything is going to happen. I would not sell this right now for any amount of money."

Saturday was a beautiful day on the mountain, warm and almost cloudless. Out at the Timberline parking lot, Swanson continued to make Geodimeter measurements. These would be the last from this observation post. The Musketeers had finally given in to the pressure and Swanson was preparing to abandon Timberline. He would relocate to Coldwater II.

For most of the day, Swanson and his Geodimeter crew hopped around the mountain in the helicopter, setting up new reflectors on the upper portion of the bulge and shooting old lines from the Timberline parking lot. The tiltmeter gang was reporting problems in some of the instruments, but there was nothing new in the ones that were working.

Jim Moore had phoned from Menlo Park to tell Swanson that new aerial maps had shown the *south* side of St. Helens had begun to bulge.

He suggested that Swanson try to establish Geodimeter lines on that side of the mountain as soon as possible.

About two-thirty Johnston was given access to the helicopter, which flew him high up the mountain. The helicopter waited with the blades turning in case they needed to get away fast. Johnston took the temperature of a new steam vent, but found it was only 190 degrees Fahrenheit, which he said was a "poor temperature" for a fumarole. At 2:42 P.M., the ground began to heave, cracks began opening in the earth, and rocks started falling. Johnston dashed to the helicopter and took off.

At Timberline, Mindy Brugman was buzzing around Swanson like a mosquito, a minor but constant annoyance. Mindy was a CalTech graduate student who had been working with the Survey's Water Resources division on a study of St. Helens's Shoestring Glacier. Mullineaux had agreed to let Mindy work in the red zone.

Earlier, before Lipman had made his discovery of the rate of the bulge growth, an anxious Don Mullineaux had contacted Mindy's boss and asked if he had an instrument that could be used to get readings on the bulge. Mullineaux was impatient with the HVO pros, who were taking too much time to get distance lines established. Mullineaux's friend said that not only did he have an instrument, but he had someone who knew it and the volcano well. Mindy arrived soon after the call with her device, an electronic distance-measuring machine called a Laser Ranger, but most of the senior (male) geologists felt that a woman should not be allowed to work on the mountain. (Sue Keiffer had worked on a nearby hilltop.) It was too dangerous for anyone to work on the mountain, Mindy was told, although male geologists had been doing just that daily.

Mindy knew St. Helens almost as well as the Denver team. She had hiked and camped on its slopes and glaciers for her research. In studying Shoestring Glacier, she had learned well how to use a Laser Ranger. The instrument was the grandchild of the Geodimeter. With the Geodimeter, the scientists were lucky to get numbers, the computations, and produce measurements within a few hours. With the Laser Ranger, Mindy was able to measure changes in a glacier minute by minute. When she saw the HVO guys were still using the Geodimeter, she couldn't stop herself from saying to Swanson, "God, I may be a moron at this, but that sure looks prehistorical."

HVO, isolated and always strapped for funds, had found and stuck with an instrument that had become an antique. Instruments like Mindy's Laser Ranger were not as precise as the Geodimeter, but they were faster and far easier to use, and for the rate the bulge was moving, they provided more than enough accuracy. Mindy had shown up with the device and after several days actually got the monitoring people to talk to her, and one by one they had fallen for the instrument. "It was revolutionary to us," said one of the HVO pros, Arnold Okamura. It didn't happen all at once. For days, the pros required measurements from both the Laser Ranger and the Geodimeter until they were certain of its accuracy.

On this Saturday, Mindy wanted to take her Laser Ranger over to the east slope and see how the warming volcano was changing her glacier. She asked to use the helicopter but, as usual, Swanson had other plans for it, and at the Timberline parking lot, he flatly refused her request. Frustrated, Mindy and a colleague, another Survey scientist from Water Resources, Carolyn Driedger, hopped into Mindy's truck and drove over to Coldwater Ridge. Dave Johnston was just setting up his equipment, including the Laser Ranger, and Mindy wanted to make sure Johnston didn't have a problem working it. Actually, she knew Johnston wouldn't. She had shown him a week earlier how to work it, and she had seen he used it easily. He, in turn, was quite impressed with the instrument. But Mindy just didn't want to leave the mountain on this beautiful day. It was exciting, and she thought Harry Glicken would be there, and he was "such an interesting creature" with something bizarre always on his mind.

Mindy's truck climbed the rough timber road to the camp. Coldwater II was little more than a ledge that dropped off steeply into the narrow valley. The road was deeply rutted and allowed speeds of only five or ten miles per hour in most places. Mindy thought that in an emergency getting in or out quickly would be impossible on these roads. Along the way, they saw cars and campers perched on other ridges. These were people who had skirted the roadblocks to get a close look if the volcano blew. Many had come for the day, but many others were there for the weekend. Driedger saw one camper on a ridge northeast of Coldwater Ridge, and her first thought was how exposed they were.

Coldwater II observation post was a dusty, ugly place. It was a roadcut surrounded by tree stumps. But it was perfectly positioned. For distance monitoring, it had an unobstructed shot at the bulge. When the two women geologists arrived, Harry Glicken was just finishing packing his truck. Dave Johnston had arrived earlier and set up his instruments and put his personal gear in the white trailer. Mindy didn't know Johnston well, although she had slept in his motel room one night because Vancouver had a shortage of rooms. But Mindy had an easy way of making friends. And Harry Glicken was funny, although his comedy was rarely intentional. Late that afternoon, they took turns sitting in Harry's director's chair, taking pictures of one another posing as if directing the next eruption.

Mindy said she wanted to spend the night at Coldwater. She said she and Carolyn had their camping gear in the truck. But Johnston encouraged Mindy to go back to Vancouver to present her case for the helicopter at the Sunday-morning science meeting. Swanson, she said, just didn't want her on St. Helens. But, Coldwater Ridge was a safe place, she said. Rocky Crandell and Dan Miller had done an analysis of the spot and found that nothing lethal had ever happened on the ridge in all of St. Helens's rambunctious past.

No, she wasn't safe here, Mindy later recalled Johnston saying.

The idea is that the mountain will fail, said Mindy, who was by now familiar with the most likely scenario. It will just go down as a slide into the valley. And the eruption will go up.

Well, Johnston told Mindy, I've found a thin layer that goes up and down the ridges here. It's a light-colored layer that's about the same thickness on all sides of the hills around here, and the only way you can explain it is as a ground-hugging blast that goes down *and* up. If that happens when you're here, it could come over the valley, up the side of this hill, and right over the top of us.

Driedger said it must be five miles from the bulge to the ridge. Surely it couldn't go that far.

Yep, said Dave, it sure could.

They talked about other volcanoes, about Johnston's adventures in Alaska, and his trip to the summit. At 7 P.M., Mindy and Carolyn said good-night, climbed into their truck, and began the drive back to Vancouver. Bouncing down the ridge road to the valley, Mindy was struck by the number and variety of animals that kept jumping into her path. Harry Glicken said good-bye to Johnston around 8 P.M.

About five and a half miles from the mountain, Dave Johnston was now alone with his volcano. Or nearly alone. In the tens of thousands of acres around him, a few dozen people were camping, nearly all of whom were there illegally. He noticed some of them in a camper on a ridge about a mile north of Coldwater II. Harry Truman was down at his lodge, maybe listening to his player piano. Scattered around other ridges, farther back than Coldwater II, were a few observers working for the state. Gerald Martin, a retired navy radioman, was in his radio-equipped camper seven miles north of the summit, overlooking Johnston's camp. At 8.4 miles, at Coldwater I, where Rick Hoblitt and a steady stream of Survey observers had spent many a cold night, was a photographer by the name of Reid Blackburn, who was working for the *Vancouver Columbian* and the *National Geographic*.[2] He was nearly two and half miles outside the red-zone boundary.[3] The rest were people who had hiked or driven around the roadblocks to camp in the hope of seeing an eruption. Overhead, an air force satellite noted the same heat increases the DOE flight had recorded, but that information, too, would not be passed on until later.

Dan Miller, in Vancouver, radioed Johnston in the old white trailer on the narrow ledge of Coldwater Ridge. Miller told Johnston he was bringing up the armored personnel carrier the next day.

"Are you serious?" asked Johnston.

"I'm serious," said Miller.

"Are they going to provide ammunition?" Johnston joked.

"That's negotiable at this point," said Miller.

Far beyond the mountain, the Spirit Lake cabin owners were preparing for the return trip to their cabins at 10 A.M. the next day.[4] Loggers, too, would be going into the area the next day. It would be Sunday, so only a small crew would be working. The armored personnel carrier was on a flatbed truck headed south toward St. Helens. Rocky Crandell was resting at home in Denver. Hoblitt, also in Denver, had just finished his manuscript and was hoping to return to St. Helens soon, perhaps the next day. Don Mullineaux was in southern California preparing to see his daughter graduate from college.

That weekend, it appeared that the increased level of concern about the mountain's new landslide dangers had convinced senior officials in the state government to expand the exclusion zone around St. Helens. An executive order to do just that sat on the governor's desk. As part of the order, a popular viewing site on the Spirit Lake Highway twelve miles west of the mountain was to be moved another four miles away from the volcano. The governor was expected to sign the order Monday.

Forest Service employee Kathy Anderson was scheduled to lead three tree-planting crews the next day on the south slope of St. Helens. Saturday evening, she discussed the plan with Dan Miller. Miller reassured Anderson that the mountain had not changed much lately. Moreover, Dave Johnston would be manning an observation post all day Sunday near the mountain, and Miller himself would be monitoring the seismic activity on the new machines in Vancouver. He said that the scientists would probably be able to give a two- or three-hour warning if things were to deteriorate.[5]

In Hawaii, a typesetter at the *Honolulu Advertiser* began keying in a story about local geologist Bob Decker, the scientist-in-charge of the Hawaiian Volcano Observatory, who had just returned from a week at St. Helens. Almost the moment he returned, a reporter called to ask if the West Coast volcano would pop soon. Decker said it could be "weeks or months" before St. Helens erupted. And when it did, Decker predicted, it would be "a very small eruption."[6]

CHAPTER 5

Swanson

J ust before sunrise on May 18, a small twin-engine reconnaissance
aircraft from the Oregon National Guard flew east to west over
Mount St. Helens. Once the Mohawk OV-1 had cleared the volcano, it
turned and flew back on a parallel course. It was 5:30 A.M. In just three
hours, St. Helens would explode in its biggest eruption in four thousand
years. Again and again, the plane flew one leg, turned back, and flew a
course slightly south of its previous path. An infrared imager in the
belly of the aircraft pointed straight down and scanned the mountain's
flanks. Any hot spots on the surface of St. Helens would be painted on
the five-inch strip of film that moved continuously through the imager.
The data from the Oregon National Guard flight would show a new line
of hot spots had appeared just beneath the surface of the bulge, but like
Friday's infrared survey, these too would not be processed before the
eruption.

The new infrared images meant magma was high within the volcano's visible cone, perhaps only a hundred yards from the surface. At
such a shallow depth, the incandescent, plastic rock was foamy and
expanding. All that was needed was one good jolt and St. Helens would
pop like a bottle of champagne.

Before the sun fully rose above the horizon, Dave Johnston pulled
his lanky body out of his sleeping bag. He may have heard the buzz of

the Oregon National Guard's plane just finishing its run. Johnston stuffed his feet into his boots and opened the door to get some fresh air. The cramped trailer had acquired a ripe odor in the weeks it had been used by Harry Glicken and other Survey observers. Standing in the doorway, with the government-issued beige Pinto station wagon parked nearby, the young geologist looked up at the clear sky and over the deep valley forest to study St. Helens.

This Sunday morning was chilly but the visibility was great. The sky was pale but clear. It would be a fine day for monitoring the mountain, and Johnston got right to work. Using Mindy Brugman's new distance measuring device, he bounced three laser shots off the bulge. He logged that the first shot was taken at 5:53 A.M. The distance between the center point of the Laser Ranger's tripod and a two-foot cluster of yellow highway reflectors was 4.80771 miles. A half hour later, the distance shortened slightly; and then a half hour after that, it grew longer again. The changes suggested the bulge had inflated slightly and then deflated, as if St. Helens was breathing.

Johnston picked up his radio and relayed his first report of the day. The signal was picked up by a Forest Service repeater located on a ridge to the north, behind him, which bounced the message south, past the mountain's peak to the Survey's operations base in Vancouver.

Bob Christiansen, who ran the Survey's monitoring program at St. Helens, recorded Johnston's data.

"What are things like up there?" asked Christiansen.

"Oh, it's very nice," said Johnston, who still fumbled with the "over and out" radio chatter used in the field. "You can see the mountain clearly. There are no clouds."

The two geologists chatted about gas readings, an interest they both shared. Then Johnston finished the morning report and fumbled the sign-off: "Okay, that's all I have to say. Vancouver, clear—I mean, Coldwater, clear."

In the Forest Service building, the small knot of scientists gathered for their daily meeting near the fire control center on the second floor and jotted down the data collected from the previous day's monitoring. Again, nothing special. There had been no eruptions in the past four days. Dave Johnston's morning deformation readings even showed some slight deceleration of the growing bulge. The tilt stations showed no consistent activity.

Seismicity was high, but no different from the rate it had been for weeks. By May 18, more than ten thousand quakes had been recorded, with twenty-four hundred of those greater than 2.4 on the Richter scale. The day before, forty-one quakes of greater than magnitude 3 had been recorded.[1] As far as Malone could tell, nearly all the quakes had originated from less than three miles beneath the north flank of the volcano, and some of them were possibly well up into the cone itself.

None of the data showed any trace of an inflection signal.

Mindy Brugman made her case for use of the helicopter and it was approved.

After the meeting, Don Swanson went downstairs to wait for someone to bring him supplies for his stay at Coldwater II. He wandered in the seismic monitoring room and heard the pens on the seismographs begin to scratch. Glancing at the machines, he saw the pens start to swing widely, in an arc as large as he had ever seen. The seismographs were recording a 5.1 quake located a thousand feet below the surface of the mountain and one mile north of the summit—just above the Timberline viewpoint.[2] The time was 8:32:11.4 A.M. PST.[3] Swanson ran upstairs to call Johnston. He wanted to know what the mountain was doing, but he couldn't raise Coldwater II.

Don Swanson raced to the local airport and hopped into a waiting Forest Service plane. As the twin-engine plane rose over an interposing hill, an eruption column appeared, spreading across the windshield and reaching as high as he could see. To Swanson, the view was eerie and marvelous. It was marvelous because he was witnessing a sight few geologists had ever seen—a stratovolcano in full eruption. But it was eerie because St. Helens just didn't look right. As they flew closer, it finally hit him: "The top of the mountain was gone. It had a flat top on it, not sharp as the day before. And when we got closer, we realized that the elevation had decreased substantially." In an instant, St. Helens had gone from the state's fifth-highest mountain to its thirtieth.[4]

Dan Miller had been driving out to the mountain to meet the Army caravan when St. Helens exploded. He did a U-turn on the interstate and slammed the accelerator to the floor. He ran inside the Forest Service building, leaped up the concrete stairs and into the communications room where he too tried to contact Dave Johnston. Normally, when anyone keyed a microphone in Vancouver, that click and an echo could be heard as the repeater picked up the signal and relayed it. But

Miller didn't hear the repeater click, which seemed odd to him because the repeater was located two miles to the north of Coldwater II, a safe distance. It struck Dan Miller then that something very large had occurred and that something terrible had happened at Coldwater II.

Soon every phone in the Forest Service's emergency communications center was ringing. Sheriffs, reporters, and rescue workers wanted to know the extent of the damage. No one in the room knew.

The first report from the field came from Kathy Anderson, who had been supervising three tree-planting crews on the south side of St. Helens. Geologists gathered around the radio as she said they were caught in a powerful eruption. Hot ash was falling heavily all around them. Their emergency route was probably gone, but they had regrouped and fought their way to an open piece of ground where a helicopter could land. She asked that her crew be evacuated. By now other reports were filtering in from television, radio, the FAA, the Weather Service, and police scanners. A major eruption was obviously under way. Anderson was told that a helicopter could not make it to her. She would have to get out as best she could, she was told. Good luck.

Swanson radioed back descriptions of what he was seeing, but he could not see much beyond the south and west slopes. The wind was blowing the ash northeast, and it made a wall in front of that portion of the mountain that was completely impenetrable. Swanson's observations were transcribed and rushed across the hall to representatives of the Oregon and Washington State Departments of Emergency Services. Washington DES had a telephone ring-down system that was obsolete and quickly failed. Information about the eruption reached many pockets of the Northwest by news broadcasts only.

In another corner of the fire-control room, the German student whose presence had kept Swanson from Coldwater II for a day began listening to the police scanners and sketching out the scene as he pieced it together from the broadcasts.

A picture was emerging of a huge eruption, perhaps the biggest ever in the lower forty-eight states. The prevailing wind, from the west, allowed observers to see that the southern half of the mountain was still pretty much in one piece, although the top had been lowered substantially. Beyond that, a large column of ash seemed to be rising from all of the northern half of the mountain, extending perhaps out beyond Spirit Lake.

The geologists knew that the vent was probably not as large as the base of the rising column of gray boiling ash. Clouds of pyroclastic flows can run along the ground to great distances. But as they move, heavier debris drops out, then ash columns begin to rise on their own thermals. So the base of an eruption column makes it appear that the vent producing the rising ash column is much larger than the vent generally is. Then again, it was possible the vent could extend from the top of the mountain to a point beyond Spirit Lake.

At Yakima in south-central Washington, the National Guard's 116th Armored Cavalry Squadron was alerted that dense ash clouds were headed their way. Knowing that their helicopters would be needed for rescue work, the soldiers scrambled to detach armaments and get the choppers in the air. Twelve didn't get off in time. They were grounded by the ash. The rest fled to the north because that was the only place they "could still see light in the sky."[5]

Malone's seismometers had picked up harmonic tremors at 11:40 A.M., tremors so strong they saturated recorders ninety miles away.[6] They would not subside for six hours.[7]

At 12:17 P.M. Swanson, who was still flying south of the volcano, observed that the color of the Plinian column had started to change from medium gray to dirty white.[8] The change in color likely signaled that the volcano had begun to release continuous, large masses of magma from the vent.[9]

Within hours after the eruption started, Hoblitt and Crandell were on a United Airlines plane. As they began their descent into Portland, they saw what had happened. Rocky Crandell was stunned.

Shortly after noon, Harry Glicken made his way to the Toutle High School on the west side of the mountain. By then, the school yard had become a staging area for helicopter search-and-rescue teams. Harry talked his way onto a chopper and directed the crew toward Coldwater II. Glicken was terrified that his friend and mentor Dave Johnston had been killed. Glicken's pilot and others familiar with the area were quickly disoriented and radioed that their maps were useless. The maps no longer matched anything seen on the ground.

At 12:44 P.M., Glicken's chopper landed at a logging camp where paramedics attempted to rescue victims in a stranded car, but, Harry Glicken wrote in his field notes, the occupants were "burned, fried, touch them and their skin falls off." He spotted an elk, standing like a statue, covered in ash, trees down all around him. The animal was dead. Facing a thick cloud of blowing ash, the helicopter crew abandoned their attempt to reach Coldwater II and returned to Toutle. At 1:15 P.M., Harry convinced another helicopter crew to search for Johnston. This one pushed closer to Coldwater II. The entire area was gray—land and air. Harry wrote, "One wouldn't know a forest existed here." The second flight was also forced to turn back, but Harry was airborne again at 2:50 P.M. While in the air, Harry heard a radio report that at Coldwater I only the top of Reid Blackburn's car was visible. The windows of the car were blown in, the photographer was dead. Glicken's third chopper was also forced to turn back. At 4:20 P.M., Glicken pleaded with an Air Force Reserve pilot to make another attempt.

"It's too dangerous," said the pilot.

"Let's try coming in from the north," Glicken pleaded.

"They are gone," he said. "Gone. Dave is gone."[10]

At 5:45 P.M., Harry wrote in his field notebook that he was back in the Forest Service building in Vancouver. He was inconsolable, overwhelmed with grief and guilt that he wasn't the one at Coldwater II.

About this time, the state police phoned Miller and told him they had blocked Interstate 5, the main north-south artery between Seattle and Portland. The Toutle River, with headwaters on St. Helens, was rising rapidly, carrying whole trees, stripped logs ready for milling, cars, and an occasional house. The police were concerned that the river would take out the Interstate 5 bridge. They asked for Miller's advice. He told them to open the highway and post observers on high ground upstream. If the observers saw a wave that could top the bridge or debris that would rip it apart, they could radio ahead and have traffic blocked.

The great billowing clouds of pulverized rock and foamy lava were being carried thousands of miles. Ash reached Missoula, Montana, at 3 P.M. and northwestern Wyoming at four.[11] Visible amounts of ash fell as far as the Great Plains.[12] The lightest ash remained in the atmosphere, circled the earth, and passed over St. Helens again in early June.[13] Some parts of eastern Washington, where the ash fall was heaviest, endured intense fallout for more than four hours.[14] A crop-dusting pilot was killed in Ellensburg, Washington, when visibility suddenly fell to zero and he flew into a power line.[15] The ash was mostly pulverized rock and lava, but it also included pine and fir chips.[16] Lighter-colored ash had several minerals derived from the volcano—including silica, potassium, zirconium, aluminum, calcium, and titanium.[17] An estimated 540,000 tons of ash fell on Washington State roads alone.[18]

By that evening, the region glowed with thirty thousand acres of burning forests, all ignited by lightning strikes from the eruption cloud.[19] The Forest Service decided it was too dangerous to fight the fires. The mudflow from the Toutle finally reached the Columbia River, where a mass of logs six hundred feet wide and twenty miles long drifted toward the Pacific.[20] Downwind, across the Northwest, hundreds of emergency vehicles were disabled by the ash and thousands of motorists were stranded.

At 8 P.M., Hoblitt got on a Forest Service observation flight. The sunlight was fading. Ash and smoke still obscured much of the area north of St. Helens. But for a brief moment, winds pushed the ash away from the summit and Hoblitt looked down in amazement. The entire north sector of the mountain seemed to be gone.

On Monday, May 19, Don Swanson awoke after a surprisingly good night's sleep. On his way to the airport, he picked up Jim Moore and Pete Lipman, who had returned late the day before. The three were in the Survey's only helicopter shortly after 7 A.M.

"That day was a glorious day for us," said Don Swanson. "It was one of those euphoric days, I think, for all of us, including the pilot. We saw evidence of tragedy, but we also saw this tremendous change to the landscape. And we figured out the grossest part of the story. We learned so much that had never been seen or known before. We were explorers."

Swanson was on his way to examine the carcass of a fresh, explosive eruption. No scientist had followed so closely the moment of such a catastrophe and lived. Swanson may have missed the revolution in geology in the sixties, but he was about to be a key player in the opening of a new era in volcanology.

The hazard mitigation group from Denver had never fully understood Swanson and the other HVO grads. The Denver hazard team was founded to develop a science that could be useful to society. The HVO grads were driven by something else. These were basic researchers whose motivations were complex and often unexamined by the scientists themselves. Ego and careerism, fame and prestige, were part of the mix, and they varied in proportion from individual to individual. But for the best of them, including Moore, Lipman, and Swanson, there was something else. The reason basic scientists do what they do may have been best explained by the physicist Richard Feynman. Feynman once pointed out that for many the motivation is simply to *find things out*.

"This is the yield," he said. "This is the gold. This is the excitement, the pay you get for all the disciplined thinking and hard work. The work is not done for the sake of an application. It is done for the excitement of what is found out . . . this is a tremendous adventure and a wild and exciting thing."

In their helicopter the volcano explorers followed the mudflows up the Toutle River, the only reliable landmark on the northwest side of St. Helens. The mountain was still steaming, and in the distance forest fires continued to burn. Swarming around the mountain were helicopters of all types searching for survivors but mainly finding bodies. Occasionally, pockets of melted ice buried in the hot ash exploded near the helicopter's path, leaving craters a hundred feet wide.

When Swanson and the others first glimpsed the north side of the volcano, they saw a landscape that had been transformed in ways that made some people dizzy. From horizon to horizon, the landscape was completely gray. Everything was seen in relief, like the figures on an ancient Roman wall. The standard reference points, features big and small, including the Timberline parking lot and Spirit Lake, were either gone or so transformed they were impossible to recognize. The scientists flew mile after mile over regions where tens of thousands of trees had been stripped and laid down, covering the ground like giant porcupine quills. The three scientists saw lots of dead animals and many that were not dead yet. They spent some time looking for Coldwater II, but none of them were familiar with its location, and it was obvious that nothing had survived so close to the volcano.

Although the day had started crystal clear, at one point the helicopter was suddenly enveloped in clouds. The pilot slowly descended until he broke through the low-lying clouds and could see the surface. They knew they were at the northern base of St. Helens and heading north, so they decided to follow the debris downhill. They flew along, tracking the debris, but oddly the altimeter showed they were not descending. They were climbing. They should have been going down into a valley, following the path of the debris. But the avalanche debris they were following was rising. Geologically, that just didn't make sense. Heavy things fall down.

Even years later when Swanson described this portion of the flight, he had a hard time convincing himself of what he had seen: "We crested out on the top of a ridge and broke through the clouds, and we realized this was the point due north of the crater at which a part of the debris avalanche had ridden up over [Coldwater] ridge. It had ridden up the saddle and gone over the ridge. It had ridden up more than one thousand feet, gone over the ridge, and flowed down the next valley to the north. And the deposit looked the same, and it was part of the deposit that filled the Toutle Valley, and yet it was going uphill."

It was a stunning testament of the power of the event. After picking up speed from the flank of Mount St. Helens, millions of tons of debris had actually flowed uphill.

Spirit Lake also confused the three geologists. It looked from the air as if it were a solid mud pot topped with timber. The three geologists speculated that the lake had been thrown out of its basin by the avalanche and the displaced water had manufactured the mudflows. Later on that day, they realized that the lake was still a lake. What they interpreted as solid mud was actually suspended sediment. The landslide had raised Spirit Lake two hundred feet higher than it was two days earlier and it had shifted northward, but it survived.

On the west side of St. Helens, the team spotted a blue car and tried to land, but the rotors kicked up too much dust. Using a hoist, a nearby National Guard Huey lowered a rescue worker, who put a flag on the car, which meant a body was inside. The body turned out to be that of a friend of Don Swanson's. Jim Fitzgerald was a geology student from the University of Idaho and had come to see the big show. The day before, Fitzgerald had been just inside the blast zone. He lived through the initial ten minutes of the blast. Footprints found in the ash showed that he had walked perhaps a hundred yards from his car. He may have realized the ash was too thick and hot to survive if he kept walking. He ran back to the car—the length of his stride was longer returning—where he died of asphyxiation. Ash was clogged solid in his throat.[21]

In all, seventeen people died of asphyxiation. Two others died when a tree fell on their tent. One person was killed in his car when a rock flew through the window. The car was parked more than eight miles from the volcano. Three people, twelve miles from the mountain, died from burns. In fact, most of the victims were cooked. Yet some people survived temperatures of up to three hundred degrees centigrade, even though their hair was singed, because the heat passed so quickly.

"We knew there had been a huge landslide or series of landslides because we saw this big hole in the mountain," Swanson recalled of that flight. "We saw how the top was removed and we saw the deposits in the valley. It took us a long time to piece this together, but by the end of the day, we had it pretty well. We knew there had been very intense magmatic activity as opposed to just being a landslide and phreatic eruptions like [what] had taken place before. Those are the most important things we learned. We also crudely put an outline around the blast area. We refined that over the next day or two, but we knew that there had not only been a blast but it devastated a very large area. That was unprecedented. All of this was a huge leap in understanding. There was a framework then that people could fit their observations into in the next few days, modify some of the things we saw, mainly fleshed out details. But the broad picture was there by the end of the day."

At dusk, after more than twelve hours in the helicopter, Swanson, Lipman, and Moore returned to Vancouver. All day, members of the Denver hazard group had been radioing for use of the helicopter. These were the people whose specialty was the interpretation of geological events from surface deposits, so arguably their observations could have been more productive. But there would be plenty of time to sort out what had happened on May 18. What was obvious was that even after the events of the preceding day, the hazard experts and the monitoring specialists still could not work together.

From Survey headquarters in Reston, Virginia, Bob Tilling ordered a comprehensive research record to document every aspect of the May 18 eruption. He asked all Survey participants to prepare papers, which would be collected into a Survey publication. A few of the HVO grads balked at the request. They wanted to publish their findings in more prestigious journals. Tilling told those few that they were free to publish anywhere they wished, but their next project grants would be decided solely on where they published. There were no further objections.

Field Notes

The eruption of St. Helens was not only one of the best-monitored eruptions, it may have been one of the best observed. Only a few of those observers were geologists. Most were people who were simply curious about the volcano and hoped to see an eruption. On May 18, hundreds did.

Within days, Survey geologists began collecting eyewitness reports, film, and photographs from people who were on or near the mountain.[1] They heard about most of these people from news stories. What these people saw, felt, and heard (or did not hear) became an extraordinary window into the heart of the maelstrom.

Early Sunday morning, flying about eleven hundred feet above the north slope, state geologists Dorothy and Keith Stoffel were in a private plane they had hired out of nearby Yakima, Washington. As the Stoffels approached the mountain, it looked "serene" at first with only minor wisps of steam coming from the summit crater. The Stoffels and their pilot had just passed over Spirit Lake and were on a course that would take them directly over the peak when they noticed a new, one-mile-long crack opening along the top of the bulge, running east-west.[2] This was precisely along the perforation line of hot spots noted in the infrared images that, at this point, were still lying on a shelf, unprocessed.

The 8:32 A.M. earthquake was felt as a solid jolt, followed by ground motion that lasted three to five seconds. Between the summit and the upper edge of the bulge, the earth rippled and churned in place for several seconds. To one observer, the north slope looked like a giant "standing wave."[3] The rolling slope kicked up dust and made the north side of the mountain look blurry for a few seconds.

Ten seconds after the earthquake had rocked the mountain, when the plane was above the northern rim of the summit crater,[4] the Stoffels saw that Floating Island, Goat Rocks, Sugar Bowl, two entire glaciers, and everything else on the north side of the mountain began "to move as one gigantic mass. The nature of the movement was eerie, like nothing we had seen before. The entire mass began to ripple and churn up, without moving laterally. Then the entire north side of the summit began sliding to the north. . . . We were watching this landslide of unbelievable proportions slide down the north side of the mountain toward Spirit Lake."[5]

The mountain released the largest landslide in human history. Enough material to bury the island of Manhattan under forty stories of trees, rocks, ice, and dirt. The slide raced downhill at over one hundred miles an hour, straight toward Spirit Lake. In less than a minute, the summer homes and lodges around the lake would be under two hundred feet of earth.

At the same time, ten miles to the east of St. Helens, at a vantage point called Bear Meadow, photographer Gary Rosenquist and his friend William Dilley were also watching the mountain. Rosenquist had just set his camera on a tripod when Dilley said the "mountain looked fuzzy." Rosenquist took a picture then turned away when Dilley yelled, "The mountain's going!" Rosenquist turned back, accidentally hitting his camera, and put the north slope in the center of the frame. It would give geologists twenty-two pictures taken over thirty seconds of a scene never before observed.[6]

Two blocks of the mountain's north face—the oblong bulge and the block between the bulge and the summit—began falling simultaneously, but at different speeds. Thirty seconds after the quake, the bulge had fallen nearly twelve hundred yards and the summit block had descended six hundred yards.[7] Approximately three cubic kilometers of St. Helens was on the move.[8] Initially, the slide moved at 100 miles an hour, but within eleven seconds it had increased to more than 150 miles an hour.

"Vancouver, Vancouver. This is it!" Dave Johnston yelled into the radio. His voice was excited, but not panicked. The message was recorded by a local ham operator but it didn't reach the USGS field headquarters in Vancouver.[9] When Johnston got no response, he tried again. His voice rising this time: "Vancouver! This is Johnston, over!"

Then something happened that would rewrite the volcano textbooks. A minute after the landslides started falling toward Spirit Lake, two explosions burst out of the mountain with an energy greater than ten million tons of TNT. A thick black jet exploded vertically from the region where the summit crater had been a little more than a minute before, and another black jet shot laterally, due north, from the ditch left by the sliding bulge block.[10]

Dave Johnston had been staring down the barrel of a gun.

In the catalog of deadly natural forces at work at St. Helens, a new one was activated at that moment—a volcano's directed blast. The explosion ripped up forests and hundreds of blocks of mountain as big as sports stadiums, dismembering them in midair, turning them into an oily black, churning cloud where temperatures reached 300 degrees Fahrenheit. All this energy was now flying over the valley and at Dave Johnston at 650 miles an hour.[11] Yet as the entire landscape was being transformed, Dave Johnston probably heard no sound. Seventy-eight seconds after the earthquake, the hurricane of stone, ice, and wood ripped Coldwater II down to the bedrock.

As the blast cloud expanded, retired Navy radioman Gerald Martin broadcast his description of the scene. Martin was working as a volunteer observer for the Washington Department of Emergency Services from a radio-equipped camper. He gave a calm description of the cloud that overran Dave Johnston's camp nearby.[12]

"Now we've got a whole great big eruption out of the crater," he called to other ham operators who were recording conversation. "And we got another opened up on the west side. The whole west side—northwest side is sliding down. Okay, boys, we got it, boys, the northwest section and north section is blowing up, coming over the ridge towards me. I'm gonna back outta here.

"Gentlemen, the, uh, camper and car sitting over to the south of me [Johnston at Coldwater II] is covered.

"It's gonna get me, too. We can't get out of here."[13] Martin was right. It did get him, too.

The Stoffels' pilot saw that the black cloud was expanding so rapidly that it would soon envelop the plane. He banked the plane sharply, pushed the throttle wide open, and dove to gain more speed. The plane was dropping at two hundred knots and still the cloud, expanding at supersonic speeds, was gaining on them.[14] In a desperate maneuver, while still diving, the pilot turned south and flew out from underneath the cloud of superheated gases, rock, and ice.

The blasts made no sound that anyone within twenty miles of the mountain heard.[15] If Harry Truman heard anything, he didn't hear it for long. Less than a minute after the eruption started, Harry and his three-story lodge were buried under two hundred feet of what had been the north slope of St. Helens.

The landslide pushed aside the water in Spirit Lake. Tons of water surged north in a 850-foot wave, and most of it sloshed back, carrying with it an entire shattered forest.[16] The surface of the new Spirit Lake was thick with uprooted trees and mud.

Light gray pyroclastic flows boiled over the mountaintop "like fumes flowing out of a beaker"[17] and down the south flank toward the Forest Service replanting crews. When supervisor Kathy Anderson saw the eruption beginning, she radioed her three crews to get into their trucks and head out over a predetermined evacuation route. One supervisor radioed back that he didn't know the route, so Anderson drove back *toward* the mountain to lead him out. As she drove, a dark cloud began flowing down the mountainside, "moving at a fantastic rate . . . throwing rocks ahead of it."[18]

When the three Forest Service trucks met, Anderson spotted a site large enough for a helicopter to land. She radioed Vancouver and asked to have her crews evacuated, by air. While waiting for an answer, a fight broke out among the workmen, who, like nearly everyone else near the exploding mountain, were close to panic. All around the Forest Service crew, strong ground winds were bending twenty-foot trees, lightning was striking the ground, and there was a strong smell of gas. A heavy ash cloud rolled over the little convoy, and Anderson ordered her crews to put on hard hats. She knew they were going to get pounded when rocks started to fall out of the cloud. Vancouver radioed back that no aircraft could safely make it in to her. She and her crews would have to

The Big Eruption: Mount St. Helens erupting at 8:22 A.M. on May 18, 1980. The bulge area slid away in a gigantic rock slide and an avalanche of debris. There was a major pumice and ash eruption. Thirteen hundred feet of the peak collapsed or blew outward. The debris avalanche filled 24 square miles, 250 square miles of land were damaged, and an estimated 200 million cubic yards of material was deposited by lahars (volcanic mudflows) into the river channels. Fifty-seven people were killed.
(PHOTO: AUSTIN POST)

drive out. The three trucks started downhill, everyone in them fearful of pyroclastic flows overtaking the convoy from behind and worried that mudflows would take out the bridges ahead of them.[19]

On the north side of the mountain, the two black eruption columns quickly combined and overtook the landslide. The blasts moved so quickly that they sheared off tree limbs and bark. Rocks were driven into the wood three inches deep as if they were bullets. And then the blast just carried away entire trees, roots and all. In large areas, the force of the blast stripped the earth down to bedrock.

Eyewitnesses saw shock waves running ahead of the blast cloud, knocking over trees as if they were dominoes. The forests were scooped up into a green wall "a mile high," said one witness, which was soon overtaken by the ash cloud. Whole stands of forests, entire glaciers, boulders, grass, and ground were now tumbling in midair.

And just as Dave Johnston had warned, the blast not only flowed down valleys but stunningly raced at more than 650 miles an hour up and over steep ridges, sending up sprays of lethal debris like black surf, which leapt over entire valleys.[20]

The wind and clouds raced over the countryside so fast they flattened campfires and held one camper's braids horizontally for fifteen seconds. One witness, driving away from the eruption at eighty to one hundred miles per hour described the cloud as a "black inky waterfall" with boiling bubbles six feet or more in diameter. Victims enveloped by the cloud felt it to be cool at first, then so blistering it burned the insides of their throats. Eyewitnesses could hear their hair sizzle and see pitch boil out of trees. Some victims had injuries that looked to doctors like microwave burns, yet their clothes were unsinged. Inside the cloud, a flashlight couldn't penetrate more than a foot. Ash, which made a gritty sound on the teeth, began filling ears and mouths.

At fifteen miles from the volcano, the cloud seemed to reach a group of terrified witnesses in seconds. Just before it would have rolled over them, it suddenly stopped and "stood up," and then the wind began rushing back *toward* the volcano at forty to fifty miles per hour.

And yet they heard no sound. The cloud was moving so fast and was so dense that sound never escaped the cloud.[21] No sound of the blast was heard inside the "zone of silence," which extended for twenty miles, but then it was heard as far away as Canada.

There was sound from the landslide, however. It was heard by James Scymansky and three other loggers cutting trees to the west of radio-man Gerald Martin's camper that morning. They heard "a horrible crashing, crunching, grinding sound" like a freight train coming through the trees. The men were protected from the full force of the blast because they were working below a ridge that stood between them and the volcano. Suddenly, it was completely dark and so hot it was almost impossible to breathe. Their mouths and throats were scalded. Each man was horribly burned. It was nearly an hour before the air cleared enough even to make out shadows. What they could see was startling. The forest around them was gone. Only pieces of trees remained, and everything was covered with ash. "It was like being put on a different planet," Scymansky recalled. It was nine hours before the four loggers were rescued. Only Scymansky survived the burns.

Fifteen miles north of St. Helens, Forest Service technician Kathy Pearson was camping with three friends when the cloud rushed at her. "All of us knew we were dead," she remembers. "It was just a matter of when it would happen because the cloud just kept getting bigger and bigger and bigger and the lightning was fierce, just unbelievable. The clouds were just purple-black. They roared and boiled and it was very, very terrifying."[22]

On the north side campers were crushed when trees rained out of the spreading cloud. The air was suddenly as hot as a kiln, scaring when inhaled, and then so filled with ash that some people had to dig it out of throats with their fingers.

Shortly after the landslide and blast, the unsupported summit of the mountain collapsed into the vent. At 8:38 A.M., a Plinian eruption column began rising from what looked to some like the entire north side of the mountain, from the summit to what was left of Spirit Lake. The rock walls of the vent that had collapsed into the crater were pulverized by the explosion and were blown sky-high.

Approaching St. Helens, a commercial pilot watched in awe as the black-gray cloud, gnarled like cauliflower, shot up to his altitude at 35,000 feet and continued to 60,000, where it flattened into a mushroom shape thirty-five miles wide. The mushroom expanded so rapidly that it overtook another airliner. Shortly, the pilot heard another commercial pilot report he was being pelted by rocks. The eruptive column and ash fall were so thick that they were blotting out sunlight completely as far as 120 miles away from the volcano.[23]

The stalk of the column, which was actually centered north of the volcano, was fifteen miles thick. The eruption column rose ten miles in six minutes.[24] Enormous bolts of lightning flashed vertically in the clouds. Pink and green lightning lit up the cloud. As the cloud passed over one eyewitness, he said balls of lightning were formed six hundred feet in the air, and then "big balls, big as a pickup . . . just started rolling across the ground and bouncing."

The ash plume contained all sorts of debris, and by 9 A.M., the material began raining out, with the heaviest stuff falling closest to the mountain and the rest carried downwind. Missoula, Montana, four hundred miles away, began getting heavy ash fall at 3 P.M.

Inside the clouds near the mountain, it became pitch-black, and people heard the debris raining around them. Breathing was difficult. Ice-cube-sized chunks of glaciers began pelting the ground followed by golf-ball-sized pellets of ash. At one point about four miles from the mountain, mud balls the size of a half-dollar fell like rain for several minutes.

The mountain was now exploding with an energy of one Hiroshima-size atomic bomb every second.[25] It would continue that output for the next nine hours.

To great distances from the mountain, the air was saturated with heat and electricity. On Mount Adams, thirty miles away, a climber was jolted with electricity when he raised his ice ax. A party of a dozen climbers, also on Mount Adams, reported they felt a rise in temperature of about forty degrees Fahrenheit that lasted for several minutes.[26]

Ten minutes after it started, the most violent aspects of the eruption—the landslide and the blast—were over. Malone's seismometers finally settled back to a level that was merely extraordinary.[27] But in those ten minutes, most of the mountain's north face had slid nearly fifteen miles. A new Spirit Lake had formed, almost two hundred feet above the old lake,[28] and its surface was thick with trees stripped of bark and branches. On the ridge below Coldwater II, the blast had circled the lower hillside, sweeping down trees in a two-hundred-degree arc. Nothing remained of Coldwater II except for shreds of aluminum from Dave Johnston's trailer scattered in the valley behind the ridge.

Trees destroyed by the blast. Over four billion board feet of usable timber, enough to build 150,000 homes, was damaged or destroyed. (PHOTO: LYN TOPINKA)

Portions of the area had been stripped down to bedrock. The body of the scientist would never be found.

The area of destruction measured twenty-three miles east to west and extended north from the mountain for eighteen miles. Almost all life disappeared within an inner blast zone nearly six miles wide.[29] About 230 square miles were severely damaged by the blast.[30]

Trees were swept away in a broad arc extending nearly seven miles north of the volcano. It was as if the entire city of San Francisco had been leveled and swept clean in ten minutes. Seven to fourteen miles beyond St. Helens, trees had been stripped of their branches and bark. Where trees had been snapped by the blasts, their stumps frequently had long splinter patterns, which splayed out like Indian war bonnets.

This "blowdown" area had two patterns. Nearest the volcano the trees had fallen in a regular radial pattern pointing away from the vol-

Rick Hoblitt excavating Dave Johnston's Coldwater II trailer. (PHOTO: C. DAN MILLER)

cano. Farther away, the fallen trees followed the topography. This second pattern indicated the blast had occasionally flowed like a river, around hillsides, in eddies.[31] This swirling left some blown-down trees pointing toward the volcano.[32]

A sample cross section of the stone-bearing wind was left on the wire cage of a bulldozer parked near one of the logging camps. The cage had been twisted and filled with the flying rocks and wood.

Beyond the fourteen-mile blowdown zone, the blast lost most of its force but was still strong enough to scorch trees, rip bark off the windward side, and drive stones an inch deep into the wood.[33] And if that weren't enough destruction, falling hot debris and lightning started forest fires farther out.

Hours after the initial landslide and blast, water began oozing from the debris and trickling downhill. The muddy water cut new paths and connected with other streams and other tributaries until it became a roaring flood of brown, soupy debris.

The largest of the mudflows originated, not high on the mountain like others in the volcano's past, but from the water-saturated landslide that filled the North Fork of the Toutle River.[34]

A wall of mud coming from the rivers on the east flank of Mount St. Helens rushed into the Swift Reservoir about a half hour after the eruption began. It raised the level of the reservoir by six feet, but because of Crandell and Mullineaux's warning, the reservoir had been drawn down thirty feet. The flow was contained and there was no damage downstream.[35]

At 10:20 A.M., a sheriff's deputy radioed that a twelve-foot wall of water, logs, and buildings was moving down the South Fork of the Toutle River. This mudflow destroyed the automatic water-level gauges, but the flows certainly topped the previous record-high flood levels of twenty-two and a half feet. At one point, the flow was measured at fifty-three feet above normal.[36]

The biggest mudflows did not begin to descend the North Fork of the Toutle River until early afternoon and didn't peak until early evening.[37] When it was over, the flows had deposited 65 million cubic yards of sediment as far down as the Columbia River, sixty miles away.[38] While the avalanche, blast, and ashfalls impacted areas to the north of the volcano, lahars ran down drainages almost completely around the mountain.[39]

Over 3 billion cubic yards of crushed rock, ash, pumice, snow, and ice filled the valley—four hundred feet thick closest to the volcano and one hundred fifty feet thick at its farthest reaches.[40] Around noon, Harry Glicken, in a helicopter searching for Dave Johnston, reported seeing numerous grayish brown mudflows moving over the avalanche.[41] These small mudflows would pond in depressions until they grew large enough to break out into the next depression, and so on, in a process that built up larger and larger flows.[42] The peak lahar did not leave the farthest reaches of the avalanche debris until about one-thirty in the afternoon.[43]

Observers were witnessing events that Rocky Crandell and Don Mullineaux had tried to work out from deposits, on Cascade volcanoes, that were hundreds and thousands of years old. Now the process was unfolding in real time.

Mudflows swept away bridge decks and girders.[44] Trees, steel beams, sections of highway, and houses all tumbled in a mixture as thick as wet concrete and rushed downstream at twenty miles per hour.[45] Steel bridges were carried downstream for miles.[46] One couple trapped by the flows saw a fully loaded logging truck floating along.[47] Then, after peaking, the flows rapidly subsided, leaving behind miles of mucky, log-jammed rivers.[48]

By the end of the day, Mount St. Helens had lost thirteen hundred feet from its peak.[49] The north side of the mountain looked as if it had been carved open, and ash clouds still seeped from the wound. The trout streams down in the valleys were buried under hundreds of feet of the mountain's north face. The once beautiful Spirit Lake valley was now a barren wasteland. Spirit Lake itself was a steaming gutter, filled with blackish water and trees. Around the lake, what few trees remained standing were stripped. Bedrock was exposed and gouged over vast areas. Because of the thick ashfall, the view in every direction was a gray monotone.

By nightfall, 130 survivors had been rescued or evacuated by helicopter.[50] Kathy Anderson and her tree-planting crews made it out safely. Harry Truman and the homes around Spirit Lake so important to their owners were buried. Fred and Margery Rollins, of Hawthorne, California, who had stopped fifteen miles from the volcano to watch the mountain erupt, were among the fifty-seven people who had died when it did.[51] More than two hundred miles of roads, fifteen miles of railway, two hundred homes, and forty-three bridges were destroyed.[52] A thousand people were left homeless, and hundreds of people who worked in the forest were suddenly without jobs. In the next few weeks, many would say that the eruption was an act of God because it came on a Sunday. If it had happened on a weekday, hundreds of loggers working in the forests would certainly have been killed.

One-quarter of the volume of the mountain was gone.[53] The volume removed from St. Helens by the landslide and blast would fill a football field to a height of six hundred miles.[54] That material was now on the once-deep, narrow valleys on the mountain's northeast side. They now looked like broad plains made lumpy by blocks of the north slope. The eruption had carved out an amphitheater one by two miles in size. What was left of the mountain was literally a shell, horseshoe-shaped.

Before May 18, 1980, pyroclastic flows had been thought to be the most devastating feature of an eruption. The volcano did produce pyroclastic flows.[55] Many of these flows were formed by the densest material collapsing out of the eruption column.[56] Generally, they ran less than five miles from the vent.[57] In totaling up the losses, both human and economic, the debris avalanche and blasts that preceded the pyroclastics, and the mudflows, were far more devastating.

Shredded in the blast was the belief held by many volcano experts that the past was the key to the future. St. Helens was one of the few volcanoes with a detailed biography already available, and that led the geologists into a false sense of familiarity. The faith they had in the past as a precise predictor of the future, which was a core belief held not just by Crandell and Mullineaux but by most volcano experts, had set the boundaries of their imaginations. They knew its forty-thousand-year eruptive history so well that they thought they knew exactly what the mountain was capable of doing. But they didn't imagine two things: the debris avalanche collapsing the mountain and the resulting powerful lateral blast that killed Dave Johnston and dozens of others.

Based on the geologic record, the Survey scientists had drawn a red zone around St. Helens. But only two of the fifty-seven deaths occurred within the red zone.

Powerful lateral blasts were never seriously considered as a possibility by most of the geologists because the volcano had only issued one in the last forty thousand years, and it was relatively minor. As Mullineaux told a Senate committee late that summer, "The lateral blast extended about three times farther than any such blast recorded by the geologic history of the volcano."[58] Also, the scientists were locked into the long-standing view of volcanic eruptions as primarily vertical explosions, classic Plinian columns. Lateral blasts were not only rare at St. Helens, the records elsewhere told geologists lateral blasts were rare everywhere.

In part, their faith should have been reaffirmed. They had correctly predicted the ash distribution, the mudflows, the floods, and the pyroclastic flows. All these closely fit the hazard maps Crandell and Mullineaux had created. But immediately following the eruption, there was widespread criticism of the Survey's failures at St. Helens, and especially of Crandell and Mullineaux's failure to call the eruption. After praising Crandell and Mullineaux for their long-range forecast, *Newsweek* magazine said the geologists' inability to predict the eruption of St. Helens was "an embarrassing failure."

Among Steve Malone's group in Seattle, there was a profound sense of depression. They, too, shared the loss of Dave Johnston, who had been a classmate at the University of Washington. That loss was magnified by the realization that the belief that had driven them to work horrendous hours for nearly two months—the belief that before an eruption a subtle signal would surely be detected if they just watched closely enough—was wrong. That was the fundamental faith they had placed in science, and that faith was now a part of the debris. St. Helens had sent no such warning.

Or maybe it had, but it was of a different sort. While many of the geologists were looking for a signal in the hours preceding an eruption, perhaps the signal was issued in late March, with the first earthquakes, small steam eruptions, and deformation of the north side of the mountain. Maybe that was the warning, and everything that followed was a tick of the geological clock.

As he had on many other days, Malone automatically agreed to make a round of television appearances on the evening of May 18, but he found himself mumbling, nearly incoherent. After one interview, he overheard one producer ask, "Why did anyone ask this guy to appear?" Back at the lab, everyone was suffering from an adrenaline hangover. They felt washed-out. The energy that had pushed them beyond their endurance week after week was gone. No one was dashing around anymore. Sentences trailed off. Why bother?

"It just felt like we were going through the motions," said Malone. "The excitement, and what amounted to hope that we are going to be able to do something significant here, was pretty much destroyed."

On the evening of May 20, Tom Casadevall returned from a field trip to Coldwater II with Dan Miller and Rick Hoblitt. Tom packed some clothes, maps, and photographs and went to Chicago to be with Dave Johnston's parents. The next day, Casadevall was in the kitchen with Tom and Alice Johnston, showing them the topographic maps of the area and the Polaroid pictures he had taken around Coldwater II the previous day. He explained that what had happened to Dave had probably happened quickly. Alice wanted to know if her son's death had been painful. No, he said. Casadevall had worked with avalanche victims, and most of the trauma that killed them came from the shock wave of compressed air that rushed ahead of the cloud.

Late that night, Tom Johnston took the exhausted young geologist up to Dave's room and said he should sleep there. When Casadevall

The "new" Mount St. Helens, with the young dome steaming. (PHOTO: LYN TOPINKA)

shut the door, he found tacked on it a quotation from Theodore Roosevelt, carefully printed in a young hand. Dave had copied it as a teenager, and Casadevall read it and then copied it himself. The quote, which applied not only to Dave but to all the Survey scientists at St. Helens, read:

> It is not the critic who counts; not the man who points out how the strong man stumbled or where the doer of deeds could have done them better. The credit belongs to the man who is actually in the arena, whose face is marred by dust and sweat and blood; who strives valiantly; who errs and comes short again and again; who knows the great enthusiasms, the great devotions; who spends himself in a worthy cause; who, at the best, knows in the end the triumph of high achievement, and who at the worst, if he fails, at least fails while daring greatly, so that his place shall

never be with those timid souls who know neither victory nor defeat.

In the months that followed, St. Helens continued steaming and shaking and occasionally even producing some relatively large eruptions. Geologists around the country began to realize that they had a unique laboratory right in their backyard, and they began laying claims to areas of research.

Pete Lipman didn't. He waited until the research teams were fully staffed and then he packed and left. To the man who had first infected Dave Johnston with a love for volcanoes, St. Helens had become an ugly place to work. Lipman left and refused to return for fifteen years, and then it was only to help dedicate the U.S. Forest Service's Dave Johnston Visitor's Center.

The eruption of Mount St. Helens had transformed many of the Survey geologists. One of them, an HVO alumnus named Dan Dzurisin, said much later, "I came to understand and appreciate that volcanoes could in fact, and I mean this at an emotional level, could in fact be very hazardous places. That was not the case in Hawaii. Hawaiian eruptions present some hazards. But all of us on the staff there wanted to participate in eruptions, wanted to get close to the vent and sample the lava flows, wanted to get as much of that experience as we could from as close as we could. At St. Helens, it really hit me at an emotional level, eventually, that these things could be dangerous, and in fact kill people, and that we had a very real responsibility for trying to mitigate the hazards. We were not only at St. Helens to satisfy our scientific curiosity. We were there to protect people and property. And that was an important part of what we learned."

Part II

The Learning Season
1980–1989

Field Notes

THE FPP EXPERIMENT

In the weeks after May 18, the geologists worked in a stupor. They were drained from the long hours week after week. They were drained by the excitement of a cataclysmic eruption. And they were drained by the death of Dave Johnston and fifty-six others. Scientists from the Centers for Disease Control, who came to study the health consequences of breathing fine ash, warned the entire Survey contingent that they were all industrial accidents waiting to happen. The CDC specialists said they wouldn't be surprised to hear sometime soon that one of the geologists had accidentally walked into a spinning helicopter rotor.

So Dan Miller decided to roast a pig. The idea took hold of him after another eruption on August 7. He would roast a pig in the steaming volcano fields at the base of St. Helens. Being a scientist meant that he would do it in a methodical fashion: notes would be kept and he would document everything. The operation needed a cover name because reporters and others were monitoring all radio communication around the volcano, so he called it the "FPP temperature experiment." FPP stood for Front Page Palmer, a name the scientists had given a local geology professor who had irritated the Survey geologists by grandstanding for the press. Miller would roast a pig and Palmer at the same time.

The Pig Experiment. (PHOTO: C. DAN MILLER)

Miller did a trial run in early August with "Chicken Little" and "His Brother." He was concerned that the sulfur odor coming off the hot ashes would add a distinctive flavor. He wrapped the two chickens in aluminum foil, dug a hole in the pyroclastic flow, buried them for half a day, and then ate the results. "They were just wonderful," remembers Miller.

He then ordered a fifty-pound suckling pig. Within a week, everything was ready. On the day of the experiment, the pig was stuffed with fresh oysters, chickens, whole lemons, huge bulbs of garlic, and fresh rosemary. Miller sewed up the pig with wire, spiced the outside, covered it up in spinach leaves, and then wrapped all that up with banana leaves flown in from the Hawaiian Volcano Observatory. The pig was then wrapped in burlap and chicken wire. Miller slid a steel bar through the wire for transport, loaded it into a helicopter, and flew out to the red zone.

The pilot, Miller, and his accomplices quickly found a cooking pit— a large explosion crater blown right through the May 18 pyroclastic flow. The helicopter landed next to the crater, and two Survey geologists carried the pig down into the pit. At the bottom, they dug a hole, tossed the pig in, poured two six-packs of beer over it, buried it, and did what they considered to be a Hawaiian chant. Before they left they inserted a thermocouple, which is generally used to measure temperatures of ash deposits, and then they flew away to do their day's other experiments. They returned at midday and found that the pig was just above room temperature. The pig was raw. Miller had dozens of people scheduled to party at 6 P.M. and he began worrying.

He organized a scouting party to find a hotter spot, although that meant going closer to the gaping ramp leading from the crater. Soon, the radio chatter was about temperatures at various spots around the pyroclastic flows. One group of scientists found a place where the temperature was 575 degrees Fahrenheit at a shallow depth, just six inches. The chief of the advance party, forgetting that communications were monitored, radioed Miller, "Dig up the fucking pig!"

Miller and his accomplices dug up the pig, put it and the iron bar on their shoulders, and began hiking out of the explosion pit. No sooner had they started than Miller, the second man in the pig parade, realized his vulnerability. All the juices—oysters, lemons, garlic, and beer— began draining from the burlap sack onto Miller's clothes and down into his boots. The pig was relocated, replanted, and by late afternoon, recovered and found to be at 204 degrees Fahrenheit, which Miller assumed meant that it was done. When the pig was loaded into the helicopter, Miller keyed the helicopter radio saying, "Vancouver, this is Five Six Yankee. We've completed the FPP temperature experiment and we're bringing home the bacon."

Twenty minutes later, the helicopter did a victory pass over the Vancouver airport as dozens of crazed scientists cheered below. It wasn't a day off, but it was close enough. The party, held in a park next to the airport, lasted late into the night. Miller kept the recipe for years, thinking he would submit it to *Gourmet* magazine, but he never did. He worried that the FPP experiment would draw criticism about government scientists wasting taxpayer dollars. He was perfectly happy in that this was probably the first time anyone had purposefully roasted a pig in hot volcanic debris, recovered it, and found it to be delicious.

CHAPTER 6

The Volcano Lab:
St. Helens After the Blast

Two days after the May 18 eruption, Rick Hoblitt, Tom Casadevall, and Dan Miller helicoptered into South Coldwater Canyon in a second helicopter Miller had arranged. Everywhere there was the distinctive odor of cooking vegetation. Chemists call it destructive distillation. It's a smell something like a mixture of rotten eggs and wet fireplace ash, and any geologist who got a whiff of it never forgot this unmistakable signature of pyroclastic flows.

In the devastated area, there was no sound of birds or wind blowing through trees, no rushing water, no electric hum of insects, only the rhythmic whap, whap, whap of search helicopters. Fifteen people were rescued from the blast zone that day, but most finds were of the dead. Some of the dead were discovered still clutching cameras.[1] Medical examiners discovered that many of the victims had their windpipes clogged solid with ash. They suffocated. A broad layer of ash, much of it only a foot deep, covered hundreds of square miles. When their helicopter landed on Coldwater Ridge, the geologists saw that in places the blast carried away tree stumps, roots, and even topsoil.

The violence appeared to be ended, although no one was then certain what remained of the mountain. Steam, opaque ash clouds, and smoke from hundreds of small fires hid most of the north face. The ash column seldom rose more than a few hundred feet above the mountain.

The depth of the earthquakes dropped from less than three miles to twenty miles beneath the surface. None of the geologists was certain if St. Helens had another blast left in her or not.

The ground was steaming. In some valley bottoms near the volcano, temperatures just a few feet below the surface were 750 degrees Fahrenheit. Occasionally, huge explosions burst from the valley floors as if artillery shells were pummeling the mud and ash landscape. These were secondary explosions. They were seeded by blocks of ice or Toutle River water trapped by the blast. More than 1.3 million cubic yards of glacial ice had been lost in the May 18 eruption.[2] Trapped in the blistering hot ash, these crushed ice blocks melted and then flashed to steam, causing their own mini-eruptions. One of the steam explosions blasted up five thousand feet, barely missing a Forest Service pilot circling the area.[3]

The North Fork of the Toutle River valley was now buried under at least one hundred fifty feet of avalanche debris.[4] In some places, it was six hundred feet deep. Vapors rose from the gray ash beds near the mountain.

The surface of Spirit Lake was completely covered with stripped trees, and its depth was only half of what it had been a week earlier. Hydrologists worried about the new lake's stability. Seeping water could undermine the muddy dam formed by the avalanche, cause it to collapse, and launch a new wall of mud toward the Columbia River.

Within days, President Jimmy Carter had arrived and toured the area with Rocky Crandell briefing him. The president was shaken by what he saw.[5] When he returned from his aerial inspection, he told reporters, "It makes the surface of the moon look like a golf course."[6]

Far beyond St. Helens, volcanic ash blanketed the region, threatening crops and power lines. The FAA warned pilots that the ash, which is as abrasive as finely crushed glass, could pit aircraft windshields and clog engines, but no major problems were reported. In Yakima, Washington, eighty miles from St. Helens, workers spent two days vacuuming seventeen tons of ash off the roof of city hall.[7]

While Swanson and the Musketeers had glimpsed a gaping hole on the north face of St. Helens on May 19, in the following days the clouds, steam, and ash kept scientists from inspecting the source of the avalanche. Then, on May 22 at 6 P.M., the dense curtain began to thin. Survey geologist David Dethier, flying with the Forest Service, saw the outlines. The spotter plane spiraled down to get a better look. Dethier saw that St. Helens had not only lost thirteen hundred feet from its peak, but the mountain was now horseshoe-shaped with a gaping opening to the north created by the lateral blast. It looked as if someone had scooped out the side of a mound of sand, leaving three high walls and an opening ramp. Jim Moore and Pete Lipman helicoptered in to take a closer look. They described the steep-walled crater as an amphitheater. At the deepest part of the crater's floor was a half ring of steaming vents surrounding the volcano's throat.

On Friday, May 23, University of Washington seismologist Steve Malone fought off the gloom that had overwhelmed his group since the eruption. He decided to replace the instrument at Elk Rock, about ten miles northwest of the new crater. In all, he had lost five seismic stations. The Elk Rock station had been planted by Malone just two days prior to the eruption. Now, exactly a week later, Malone was dropped off by helicopter in a spot that bore no resemblance to the previously wild, beautiful country.

"It was spooky, so totally unfamiliar in an almost scary way," he said of his first view of the devastation. "Here was this place we'd been one week before, exactly, and it had been green and lush. And now that was all gone. Nothing was left but the coating of ash and bits of the station that were bashed and battered and melted. I was pretty awestruck."

When he returned to Seattle, he saw that, for the first time in almost two months, no earthquake greater than a 3.0 had been recorded in his absence.

Two days later, Malone's reestablished Elk Rock station was disabled by St. Helens's second eruption, which, compared to that of May 18, Hoblitt described as a "moose fart." At 2:30 A.M. on May 25, a column of ash exploded out of the crater. It reached forty-five thousand feet and in minutes formed into a mushroom cloud.[8] Dick Stoiber, a volcanic-gas specialist from Dartmouth College, measured the highest levels of sulfur dioxide yet recorded at St. Helens.[9] Soon, Dave Johnston's partner, Thomas Casadevall, discovered that St. Helens was emitting at least as much hydrogen sulfide as it was sulfur dioxide. It was as if the gas jets that had been closed in the weeks before the blast were now thrown wide open.

For most of the next day, St. Helens continued to eject ash, which mixed with a storm from the Pacific. "It was raining mud," said one observer. The mud rain slathered highways, causing one fifteen-vehicle pileup and countless other cars to slide off roads. Busloads of National Guard troops, on their way to help with the cleanup at St. Helens, were stranded thirty miles away.

Malone's Elk Rock station wasn't fixed until June 10. Two days later, the mountain erupted again.

St. Helens's growing reputation as the most attractive geology lab in America was creating a rush of eager scientists to the site. The Survey had a more orderly process. A meeting was held in Vancouver in which various scientists were assigned different aspects of the eruption Hoblitt and Miller and a few others got the blast.

Working in the volcano lab was a challenge. It was a hot, dangerous, stinking place that was always threatening to explode. The thick pyroclastic flows that lay over the ramp were a lot like fireplace ash, except they were hot enough to roast a pig—or a groggy geologist who stumbled into the flow. The pyroclastic flows had not had centuries to degas and compact into the debris layers Crandell and Mullineaux had examined. If you threw a rock into one, the missile landed with a splash, as if it were water, and the rock would slowly disappear beneath what had looked to be a solid surface. One scientist slipped into the ash. She was yanked out almost instantly but still suffered third-degree burns over most of her legs.

Hoblitt wanted to measure the temperatures across the pyroclastic flows, but first he had to figure out a way to walk across them. He discovered that if he stepped slowly, wiggling each foot carefully down into the material, he could slowly release the trapped gases enough to compress the ash and create a solid foothold. It worked fine as long as he was patient, didn't lose his balance in the rubble, or didn't need to run from an eruption.

Danger littered the St. Helens lab. If it wasn't an eruption from the vent of the volcano, there were the steam explosions in the valleys, some of which left craters big enough to fly a helicopter into. While working on the flows, geologists worried that they were walking over one of these land mines, or that a new pyroclastic flow would come pouring over the lip of the mountain's crater and overtake them like a flash flood.

Back at the Survey's headquarters in Reston, Virginia, officials debated whether to establish a permanent volcano observatory at St. Helens. Many of the geologists working at the volcano pressed hard for a permanent observatory. The summer's destructive eruptions were giving way to eruptions that were constructing a series of small lava domes in the crater. Those arguing for a permanent observatory said that St. Helens could provide insight into how volcanoes explode and perhaps eventually offer a unique glimpse of mountain building. Also, fresh deposits were spread all around the north face of the mountain. At last, scientists could examine in detail processes about which they had only guessed. With these observations, they might be able to better interpret the old, cold deposits they were finding around other volcanoes.

Moreover, St. Helens was also producing streams of new data—on earthquakes, gases, tilt, and ground deformation—every minute. Perhaps in the flood of data pouring into Vancouver every day was the signal that St. Helens was preparing for another eruption. Eruptions had been predicted before, in Russia and Japan, but without a high level of accuracy. The Survey had an opportunity at St. Helens to learn how to read an explosive volcano's behavior and perhaps learn how to predict, with confidence, a coming eruption.

Those opposed to a permanent observatory argued that St. Helens could stop erupting at any time, and the Survey would be saddled with a expensive outpost of marginal value. But by the end of the summer, the issue had been decided. The Survey would create the David A. Johnston Cascades Volcano Observatory, CVO. The name itself was meant to regionalize responsibilities beyond St. Helens to include other West Coast volcanoes such as Mammoth and Shasta in California, Hood in Oregon, and Rainier and Baker in Washington State. Also, officials argued there were four thousand volcanoes in the Cascades (if you count every cinder cone), and it would be best to preemptively gather baseline data. CVO would also be a launching pad to respond to the most violent volcanoes in America, those in Alaska.

Don Peterson, an HVO alumnus, was chosen to be CVO's scientist-in-charge. One of Peterson's first moves was to get Don Swanson to pledge to move to CVO permanently. Peterson was also given a budget that allowed him to hire a few of the promising young "part-time" geologists who had essentially been working full-time for the Survey for years, including Tom Casadevall and Dan Dzurisin.

Harry Glicken, Dave Johnston's idiosyncratic assistant, had a driving ambition to become a Survey geologist, but he was not invited to join. He was still a student, years away from earning his Ph.D. The Survey even left him out of the autopsy assignments.

In Seattle, Steve Malone's band of seismologists at the University of Washington combed through the pre–May 18 seismic data again and again. They looked to see if they had missed something, anything that might have warned them of the approaching disaster hours or a day or two before. They found nothing.

To Malone and many in his group, it seemed that their work was nearly useless. So, Malone decided he needed to look at St. Helens as a new volcano.

"We had to start from scratch and learn a whole new way of thinking about St. Helens," he said years later. "Besides, all the stuff that had happened before May 18 had led to squat as far as being helpful. So we started over, and that's what it felt like."

Throughout the summer, the volcano made a lot of seismic noise. Large rockfalls split off the crater walls and tumbled onto the floor. And even the steam explosions rocked the seismometers.

The quakes Malone's group began seeing were small, in the 1.0 to 3.0 range, not the 4.5 and 5.0s that had rocked the mountain before. Malone began to realize that he was calling these new quakes "small" by comparison to those of April and early May. At any other time, they would have made him curious. So part of the new way of looking at St. Helens quakes was to realize that the energy being released was significantly less, but still significant.

Buried in the new seismic data, Malone and his gang began to see a signal. Examining the data collected prior to the May 25 eruption, they saw what appeared to be harmonic tremor. It was fairly subtle, a fairly small pattern, right down in the noise. They began looking for the pattern in the incoming data and saw it again. For several days, Malone's group monitored low-amplitude harmonic tremor. Then on Monday night, June 2, it disappeared. At 2 A.M., June 3, harmonic tremor reappeared and began increasing strongly. Malone phoned Vancouver and warned them that the pattern his group was seeing was like that found in the data prior to the May 25 eruption.[10] Logging was shut down and scientists were pulled from the field, but there was no eruption. Malone's call was a false alarm. The next day, the red zone was reopened for loggers, geologists, and the news media.

On the afternoon of June 12, Malone saw the same signal, only this time it was a little cleaner. The low levels of harmonic tremor increased in late afternoon. Malone called Vancouver again and Vancouver notified the Washington Department of Emergency Services, the Forest Service, and Weyerhaeuser that some activity was expected "imminently," remembers Dan Dzurisin, who was scientist-in-charge on that day. Within two hours of Malone's call, at 7:05 P.M., St. Helens popped an eruption cloud to thirteen thousand feet. Immediately after the eruption, the tremor stopped. Within the next hour, the tremor picked up again, and at 9:11 P.M., an explosion occurred, this time sending a cloud to thirty-five thousand feet and pyroclastic flows pouring down toward Spirit Lake.

"This was a big boost," Malone remembers. "Again, we said, 'Okay, maybe there is some purpose to our lives.' We were collecting interesting data, digital data, the first time this type of thing had been recorded this way. We've got this amazing data set. That sort of thing was going on. But when you can turn around and say, in real time, here's a warning, and you make a difference, that's when it becomes easier to become emotionally involved. The objective nature of the scientific work changes because you can actually have an impact."

In Vancouver, Malone's newfound skill of predicting eruptions took some time to get used to. The HVO pros had believed that calling an explosive eruption within hours was like winning the lottery. The odds were very long. Everybody knew that. Malone's June 3 false alarm didn't help his credibility either. And then the Survey had not established an efficient communication link to Malone. Sometimes, key Survey officials just didn't get the word. Malone once telephoned a warning of an eruption but the Survey geologist who answered the call was a new hire. He had been on duty in Vancouver just a few days, and he failed to pass Malone's message along to his supervisors. Sure enough, the volcano blew. Another time, Malone could not get anyone from the Survey to leave a planning meeting to take his call. So he phoned the Forest Service and told them that St. Helens would soon erupt. And it did.

By the end of the summer, however, people were coming to believe that quake patterns might reliably predict eruptions, at least St. Helens's dome-building, dome-bursting eruptions. The information became especially valuable to the Survey as geologists began working in the gaping horseshoe-shaped amphitheater that once was the heart of Mount St. Helens.

Working in the crater was even more challenging and dangerous than working in the pyroclastic flow plains. First, the crater was an efficient fog generator. The cold, stagnant air at six thousand feet mixed in the amphitheater with the humid sulfur fumes rising from beneath the crater floor. Visibility could drop in minutes so that geologists working in the crater couldn't see their boots. When they were thus blinded, rockfalls reverberated around the amphitheater, sounding as if landslides were coming from every direction.[11]

Another problem in working in the crater was that its floor was often split by lava, and walking into incandescent rock was not advised. Even standing still could be dangerous. Large rock avalanches split off the walls of the crater and tumbled halfway across the floor. At other times, the growing dome blew boulders—some were of softball size and others were as big as cars—which rained down inside the crater. Geologists always began their workday by scouting for hiding places they could duck into in a hurry.

Even commuting to work was dangerous. More than one geologist thought the most dangerous aspect of working in the crater was flying through turbulent thermals and around the volcano in those flimsy helicopters. The wind shear at the crater's ramp, on the north side, was strong enough to whip a helicopter around like a leaf. Twice, the wind flipped choppers over. On rare occasions, the Vietnam vets who had been hired to fly the geologists refused to return to the crater or were unable to return because of the weather, which sometimes meant the geologists had to hike out across the field of debris.

The biggest hazard facing geologists working in the crater was a gas or phreatic explosion. This kind of "minor" event, probably affecting only those in the crater, could certainly be lethal.

Given all these hazards, a debate arose over the value of working in the crater at all. As in other disagreements they had had in the past, the field geologists' scientific debate could devolve into verbal mud wrestling. And again, it was the Denver hazard assessment group, bolstered by a new hire named Chris Newhall, arguing that the specific benefit fell far short of overriding the real risks. In addition to the risk/benefit argument, there was real concern that if another geologist were killed, it could permanently cripple the entire CVO research program and perhaps even the Survey itself.

The geologists on the safety-first side included Chris Newhall and another new hire named John Dvorak. Both men argued vigorously against working in the crater. Eventually, as the monitoring continued close-in, both men began feeling they were being isolated.

"People thought either I was scared or overly cautious or I was putting my nose in other people's business," said Dvorak, who eventually left the Survey. "It got to the point where I got a big black eye out of it, and it followed me around for my career."

The argument over personal risk and scientific reward lasted months. Don Swanson led the group pushing to work as close as possible, just as he had prior to May 18. It was the way they had worked in Hawaii. He and the other HVO grads, including the scientist-in-charge at CVO, believed this was the most effective way to work on St. Helens. Swanson wasn't about to force anyone to go into the crater, but he didn't want to be forcibly kept out of a rich work environment. He didn't exactly win the argument. He just kept arguing until everyone else was exhausted by the subject. Eventually, the press began calling him "Mr. Crater." Later, in a tribute to another geologist killed in an eruption, Swanson laid out this thinking about taking risks for their science:

"It is well to remember that volcanology is the study of volcanoes, and that purely scientific, curiosity-driven motives are as justified as those designed purely to mitigate risks, and I think more valuable in the end," Swanson wrote. "Curiosity leads to understanding, and understanding is the paramount goal of the science as well as the soundest basis for reducing risk. Volcanologists who are curious will get themselves into trouble and sometimes die because of it. It is often stated that we must weigh the potential benefits and risks before doing something that may be perceived as risky. Of course we must, but it is mathematically impossible to solve one equation with two unknowns, and generally the potential benefits and risks are both unknowns. In the end, it comes down to common sense, which varies among individuals and in any case is far from foolproof. Let it be no other way, and let us praise the curious as we mourn the dead."[12]

Soon the seismic tracings were allowing geologists to produce a nearly perfect string of predictions that lowered much of the risk from dome eruptions. But there were still great dangers from avalanches and minor explosions.

Swanson and others were eager to study the gestation of the volcano's dome. Dome building is a common part of many eruptions and it can go on for years. But studying dome building up close was dangerous at most volcanoes. The domes that are built and blown apart are made from thick, nonexplosive magma. Like toothpaste, domes are squeezed out onto the surface where they cool and turn to rock. The toothpaste tubes extend down toward the magma plug that feeds the growing dome. Young domes are always growing. They are either being squeezed out or expanded internally by rising magma. Normally, the weight of the dome suppresses the magma until the upward force is greater than the overlying weight of the dome. At that point, the dome can shatter in an explosion, blasting new magma and old rock.

Sometimes, gravity alone can cause a segment of the dome to collapse even without an eruption. This can be lethal as well, because as the dome collapses, it can spawn pyroclastic flows. Dome eruptions in Japanese and Indonesian volcanoes had killed thousands of people in pyroclastic flows. So learning about dome building and collapse was a high priority for volcanologists everywhere.

St. Helens had plenty of old lava domes. The pre-eruption summit had been filled by a dome four hundred to six hundred years old.[13] But some of the volcano's domes on the north face were now ash in Yakima or they were part of the debris field that blanketed the bottom of the North Toutle Valley.

The first installment of St. Helens's first new lava dome erupted on June 12, 1980, but it wasn't until June 15 that the weather cleared enough for observers to see it. The dome was a gray, gnarled rock wart, 700 feet wide and rising 130 feet off the relatively flat crater floor. The dome's surface was broken, like crusty bread, and in the cracks geologists could see the slight red glow of incandescent lava.[14] It was the first lava spotted. By June 27, the dome had grown to be 1,200 feet across and 200 feet high.[15] This dome was almost completely blown away in the July 22 eruption.[16] Eventually, the replacement dome grew to be a half mile wide and 1,150 feet tall.

Many of St. Helens's eruptions in the 1980s were dome-building events. For a time, the volcano would squeeze up some lava that would form a blister on the crater floor. The mound would harden even as more magma was pushing up from below the new dome. Eventually, the pressure from below was greater than the pressure of the overlying rocks, and *boom*, rocks—red, black, and gray—exploded.

Dome building, like many other aspects of stratovolcanic activity, was rarely available for scientists to study. Many explosive volcanoes were in remote and relatively inaccessible regions of the world, such as the Kamchatka Peninsula, which at the height of the Cold War was off-limits to U.S. scientists.

Despite the obstacles, St. Helens was the best lab for studying dome growth. It was easily accessible. Platoons of geologists were available. And most important, Steve Malone's warnings provided a significant new measure of safety.

Swanson, still a deformation man, began his research on dome growth by concentrating on the cracks that radiated from the dome across the crater floor. In typical fashion, Swanson often used simple tools from the hardware store. He would pound masonry nails on either side of a crack and measure the distance between them with a steel rule—often over a vein of incandescent rock three to six feet wide. He soon found that the number, size, and most important, the rate at which the cracks deformed were good predictors of eruptions. When the tip of the rising magma moved up underneath the dome, the cracks would widen. Eventually, the pressure beneath the dome increased to instability. Soon magma would enter the dome and a new lobe of lava would blossom from the dome, or later when the dome was large, it would cause the dome to expand like a balloon.

The work was tricky enough in the summers, when the fog would blind the geologists for long periods. But during the winters, it was almost foolhardy. Then, hot gases from the cracks along the crater floor melted snow. As snow piled up, the warm cracks produced snow caves. Geologists working on the snowy crater floor then had to be careful they didn't break through the roof of one of these snow caves and fall into the liquid rock and hot gases. Swanson, however, went further. He would descend into the snow caves and crawl along the slippery ledges of the snow tunnels to get his measurements.

"You are looking down a crack at incandescent rocks with no way out because you're in a snow cave," says one person who regularly accompanied Swanson on his rounds. "If you slipped off the ledge, you're dead. That was probably one of the more risky things I've ever done. It wasn't a macho thing, because Don would have done it by himself where no one knew about it. But he had spent so much time measuring those cracks, and he wanted to make sure we could continue to have that data before every eruption."

Swanson had, at last, found an inflection point.

Beyond the crater, St. Helens was proving to be a volcanologist's dream. For example, geologist Norm MacLeod had spent years puzzling over a strange deposit in southwestern Oregon—the site of Mount Mazama. About 5000 B.C., Mazama exploded in an eruption that was ten times larger than Vesuvius's A.D. 79 eruption, more than twice as large as Krakatau's 1883 eruption, and forty times larger than St. Helens's in 1980. So much material was ejected that the overlying mountain collapsed into the magma chamber below, leaving a depression that was eventually filled by rain to form what is today Crater Lake.

MacLeod had been working on the pyroclastic deposits around Crater Lake when he saw one that completely stumped him. It was a thin, pinkish layer, separated from the main pyroclastic deposits by a good distance. Trekking around St. Helens's debris, MacLeod discovered what the Crater Lake deposit really was: a secondary pyroclastic flow. When the energy of these great searing winds ran low, and it dropped the heavier rubble, the hot, turbulent air kept aloft fine particles that could still travel a great distance down even slight slopes. Eventually they settled, too, compacting into thin layers, and moisture would oxidize the iron in the dust, turning it pink.

"That wasn't the only example," says MacLeod "It was a daily event, seeing things you'd seen before but not realizing what they were. And a lot of times it was not that you could not figure out what it was, but you didn't ask the question. That was the main thing. I didn't ask the question of 'Why is this pink ash where it is?' At St. Helens, I saw what it was."

Over several years of looking at the deposits around St. Helens, MacLeod saw what time soon did to them. Compressed by rain and churned by gophers and other animals, these signs of lethally hot pyroclastic flows that had once seared everything in their path began to vanish.

"A lot of the stuff I studied in the eighties is no longer there," said MacLeod. "The little deposits are very ephemeral things. They are wiped out almost immediately. I come back now and I can hardly find the pyroclastic flows, and yet they were very common there."

Learning that the traces of once-lethal flows can be ephemeral put new light on the differences Dave Johnston and Rocky Crandell had had interpreting the geologic record. Johnston was concerned that Crandell and Mullineaux's hazard maps were too conservative, and that the thin layers he had come upon indicated the dangers extended far beyond the red zones drawn on their maps. In the debris was evidence to settle differences Johnston had had with Crandell.

"Crandell and Mullineaux, not having a lot of experience on active volcanoes and just looking at the history of the volcanoes, can only go with the deposits they see," said MacLeod, who went on to become the scientist-in-charge at CVO. "Sure, they underestimated the extent of things. Dave Johnston came from Augustine. He could see young deposits. Everybody comes with different eyes."

Lying in pieces around St. Helens was also the mechanism for the explosion. Rick Hoblitt actually found a piece of it, which he kept on his desk. It's a piece of rock or, more precisely, it was two pieces of rock welded together.

What had led to the blast of May 18 was a feature known as a cryptodome. Essentially, magma had pushed its way high into the cone of the volcano and was blocked. Escaping gases made the magma lose its buoyancy, allowing the crystals in the magma to grow, giving the lava the consistency of peanut butter, or as geologists say, making it more viscous. As the rising plume of magma stopped, its leading edge began to crust. Hoblitt's souvenir was from the meeting of the old and new rock.

The outer, hard crust and the mountain above it pushed down on the magma with a force of more than thirty-one thousand pounds per square inch.[17] But the lower part of the magma column was still buoyant and wanted to rise and exerted increasing upward pressure. It was like two elephants standing on a seesaw. One, however, had slightly more weight than the other. It pushed out on the mountain's weakest spot. So the mountain began to bulge. Then an earthquake cracked the cryptodome.

In the summer of 1980, Harry Glicken returned to St. Helens. He wanted to be a part of the autopsy teams. He was driven to detail the lateral blast that had killed Johnston. But by then, every aspect of the eruption had been claimed by different Survey scientists. Moreover, Glicken was just a doctoral student at UC Santa Barbara, studying with the well-respected Richard Fisher. But Glicken was not a Survey employee. And he was probably hobbled most by his well-deserved reputation as a nut.

"Harry was a character his whole life," recalled a friend from the St. Helens days. "Everyone who knew him was absolutely amazed he was such a good scientist. You couldn't believe that a person who was so disorganized and has some of the personality traits he had was such a good scientist.

"Harry was a terror to drive with. His attention to what was going on around him was almost nonexistent. He would drive at full speed down the road, talking about whatever was important to him, and we'd come to a four-way stoplight and he'd sail through it, never knowing he'd just gone through, sending drivers up telephone poles in his wake. He was a cartoon character. If you ever rode with Harry in a car once, you never allowed him at the wheel again with you in it. Never. He was a little wiry guy. Extremely sensitive. A strange guy. But if you sat down with him, you'd just be knocked out about his mind. As far as where he was going—he had no clue. But he was extremely engaged in detail. He just saw everything and described everything. He was the kind of guy you would not think could give a talk in a professional meeting, but he'd get up there and he'd give an excellent organized talk and sit there and defend himself with the best."

Disappointed that he had been left out of the catastrophic eruption autopsy, Glicken was wandering the temporary observatory headquarters in Vancouver when he bumped into Barry Voight, the landslide expert. There had been a new appreciation of Voight's landslide paper after the eruption. Voight was given an adjunct faculty appointment to the Survey and asked to head the investigation of the avalanche. Glicken asked Voight if he could join the avalanche team. Voight thought Glicken was eccentric but a solid scientist, so he hired him.

"The Glick," as he liked to be called, tore into the work. Many of the Survey scientists working on the volcano at the time interpreted Glicken's fever pitch at work as survivor's guilt. Indeed, Glicken was anguished by the death of the man he considered his first real mentor. To family and friends, he said that St. Helens should have taken him and not Johnston because Johnston would surely have made outstanding contributions to the science.

Guilt was not the only thing driving Glicken. After a while, a new passion motivated him. Harry Glicken wanted to be a member of the Survey. During the brief period after the May 18 eruption when the Survey was hiring young geologists, Glicken, without his Ph.D., was not considered. So he planned to do the best possible work on the landslide, in tribute to Dave Johnston and to demonstrate what kind of contribution he could make to the Survey.

It's hard to imagine a job more daunting. About one-quarter of the volcano had fallen in the avalanche. Some of the intact sections were a hundred yards wide, while other pieces had been shattered into thousands of fragments. When the avalanche came to rest, the debris field looked like a vast range of miniature mountains, called hummocks by geologists. Some of them stood two hundred feet high. This chaotic, bumpy landscape spread for miles beyond the old volcano and into the Toutle River valley.

Harry and the team he worked with put those pieces back together again. Not literally, of course, but they did map the debris field in meticulous detail. Then they traced back sections of the deposits to show where they had originated, how they had traveled, and why they came to rest where they did. It was an outstanding piece of work that impressed many of those who knew Glicken.

Swanson acknowledged that Harry's work was "the most complete study of such a deposit" that he had ever seen. In Swanson's estimation, Glicken had quickly become "a world leader in studies of volcanic debris avalanches."[18] And soon Glicken began traveling—to Indonesia, New Zealand, Guadeloupe, and Japan—to study deposits and lecture on their significance.

Harry and the other two members of the team found, buried in the avalanche, something about tall volcanoes, as opposed to the relatively flat shield volcanoes, that had not concerned volcanologists much. And that is that tall volcanoes collapse. Their avalanches had been seen before, for example, at Bandai-san in Japan. But it was a little-understood and vastly underappreciated characteristic of steep-sided volcanoes. Once the St. Helens hummocks had been recognized for what they were—evidence of a collapsed volcano—geologists began spotting them around the world. More than two hundred such hummock fields have been identified, from the West Indies to Hawaii. In fact, geologists now believe that more than half of the island of Molokai in Hawaii was carried away in a submarine avalanche.[19]

No one before had correctly figured out how these hummocks were created. Many geologists had speculated that they were remnants of mud-flows, trails from retreating glaciers, or explosive deposits. Hummock fields began to be noticed nearly everywhere geologists looked at volcanoes. The Castle Crags Chaos in California, for example, covers an area larger than the city of Los Angeles and contains blocks up to a mile long.

St. Helens demonstrated that even volcanoes cannot escape the force of gravity. They will grow only so high and then they will fall. They can fall in an eruption, as St. Helens had, or they can fall without an eruption.

It also meant that to the list of volcano danger—from mudflows to pyroclastic flows—a new danger had to be added: the debris avalanche.

A few years after Glicken started his work, it was becoming clear that the study on the avalanche was unique, thorough, and opened new vistas for volcanology. And it was also becoming clear to everyone at the Survey that Harry Glicken wouldn't be hired. He was not going to realize his dream. The reasons were complex. First, while the Survey budget for volcano programs increased after the 1980 eruption, by the mid-1980s, when Glicken was finishing his work, things had changed. St. Helens was clearly moving toward the end of its eruptive cycle. Dome-building eruptions were slowing and would stop altogether in 1986. Then, the Survey geologists at the Cascades Volcano Observatory would begin to examine whether CVO should close down when St. Helens had gone quiet, perhaps for another century.

But The Glick had another obstacle to overcome—himself. He had a unique personality. Harry Glicken was different. Very different. He was chatty, disorganized, bubbling with ideas, and generally lost in the real world. Many people, especially a few young women geologists, thought Glicken was attractively eccentric. Many senior Survey geologists, all male, thought he was a space cadet. Actually, Harry Glicken was in a class by himself. "Harry was Harry," says his sister.

"Harry was very enthusiastic, very bright, and very ambitious, ambitious to do something worthwhile on volcanoes," says Robin Holcomb, a friend from those CVO days. "But he was having trouble finding a place for himself."

Looking downstream at the North Fork Toutle River valley, north and west of Mount St. Helens, showing part of the nearly three cubic kilometers of debris avalanche that slid from the volcano at the May 18 eruption. The avalanche traveled approximately 15 miles downstream at more than 150 miles per hour. It left behind a hummocky deposit that varied in thickness from 150 to 600 feet. (PHOTO: LYN TOPINKA)

Harry used his apartment's smoke alarm as a baking timer. Fish was finished when smoke filled the room. At college, a professor who gave Harry the use of his office returned weeks later to discover Harry was sleeping there and had left parts of take-out hamburgers scattered among the debris. Harry once got off a bus and watched it disappear before he realized he was barefoot and his boots were now moving down the road.

The Glick began to realize that he would never be a member of the Survey, and that realization pushed him into a depression. Associates noticed that Harry was pulling the hair out of his head. Glicken began wandering the world, an itinerant volcano-landslide expert, working on grants and fellowships. Eventually, he seemed to find a place for himself in Japan, teaching, doing research, and to pick up some spare money, acting as a translator.

· · ·

By the end of December 1982, St. Helens had had fifteen eruptions, most of those predicted within a few tens of minutes to a few hours. The last seven, starting in mid-April 1981, had been predicted between three days to three weeks in advance. In a paper written for the journal *Science*, Don Swanson and others said that the "single most important source of data for short-term predictions" is earthquake data. But he also said that deformation of the crater floor and the lava dome, as well as gas emissions, also provided data for the predictions.[20]

The seismologists and geologists learned that eruptions were preceded by shallow earthquakes. Those quakes could begin from a few days to two weeks before an eruption. Then the energy released by earthquakes began a sudden upward curve a few hours before an eruption. Electronic tiltmeters placed in the crater also recorded the ground deforming outward from the dome prior to an eruption. Swanson had recorded new cracks—from hairline fractures to some eleven feet wide and thirty-two feet deep, running away from the dome—in the crater floor several days to four weeks before an eruption, and these cracks widened just before the eruption.

On March 12, 1982, earthquake data and deformation of the dome and crater floor triggered a warning that "an eruption is likely within the next three weeks." On March 15, the deformation greatly increased, and at 7 P.M. the warning was revised to state an eruption "will probably begin within one to five days." Another revision issued at 9 A.M. on March 19 set a window of twenty-four hours. A little over ten hours later, the south side of the dome blew out blocks of hot pumice into the snow-covered crater wall, melting snow and triggering a flood that reached the Toutle River valley.[21]

St. Helens had other important lessons to teach. A plane flying south of the volcano at 7:30 P.M. on March 19, 1982, saw glowing objects shooting out of the crater. It was just another explosive eruption and, on the scale of things, rather puny. The eruption materials, mostly ash and pumice, never reached much beyond the crater itself.

If it had been summer, the eruption would have been recorded as a minor example of dome building. But it was winter, and the crater floor and walls were covered in snow.

The small spray of pumice, hot gases, and possibly steam melted much of the snow inside the crater and created a small lake. Within minutes, the lake was large enough to surround the dome, which was then about six hundred feet high. Water began pouring through the mountain breach. The warm water cut a ravine ninety feet deep and began to cascade down the mountain at about twenty feet per second (13.6 mph). As it descended, the rapidly moving floodwaters picked up debris from previous eruptions, turning the mass into a mudflow. The flow divided. One arm dumped into Spirit Lake. But the other arm raced across the pyroclastic-flow deposits, reached the Toutle River valley, and breached a retention dam twenty miles from the crater. A survey of the mud lines confirmed that the Toutle River had risen to within a few yards of its highest levels, in the May 18, 1980, mudflow. The minor eruption had triggered a large and dangerous flood because it had spewed its hot material on deep snow.

The eruption generated a paper that appeared in the journal *Science*. And then it was largely forgotten. But a similar eruption would happen soon and would be impossible to forget.

Even with their growing success, the geologists were uncertain whether their newfound predictive powers would work at other volcanoes. Generally, they thought they would not.

"Experience is crucial," wrote Swanson and others in a 1983 paper in the journal *Science*. "We must be able to observe repeated episodes if we are to acquire knowledge of causes and to gain confidence in our predictions. Only rarely will an initial episode have such an obvious cause that the course of coming events can be correctly predicted."

At the peak of the dome eruptions, CVO was overwhelmed by data—gas data, seismic data, tilt data, temperature data, magnetic data, data, data, data. They would be reported at every science meeting. The gas guys would give their readings. Then the deformation team would do their act. And on and on. They would occasionally try to integrate their data, as Jaggar had done at HVO, but that took extra time and effort.

Then Tom Murray, a bear of a man who was called the biggest volcanologist in the world, began doing some tinkering. He was renowned as a computer and electronics genius. Murray once submitted a research grant application to the Survey with just two words: "Trust me." It was funded.

Murray began wondering if he couldn't design a tool to weave the different streams of data into a single coherent picture. Perhaps gas readings were related to deformation, and seismic levels to tilts. Perhaps there was a way of doing comparisons quickly that would display changes over time. Murray created a software program called BOB, which did all these things, and it changed how scientists looked at volcanoes.

"It wasn't really until St. Helens that we really developed an organic view," says CVO scientist John Ewert. "BOB allowed everybody to put their data into the same system and look at the data on the same time base, at the same scale. All of a sudden, you see all these things happening together, and you can look at it in almost real time. You can see how these things are changing, hour to hour, or certainly day to day. When you get everybody throwing their observations and their data into the same pot, and you mix it all up and then you look at it, you start to see how all the little pieces start to fit together. That is one of the big changes between how things were approached pre–St. Helens and now."

BOB also helped integrate more than data. It helped build teams.

"Everybody was doing their own thing," says Ewert. "St. Helens changed a lot of that. And the advent of the BOB system and getting data onto the same common time frame was a major step toward the teamwork aspect of this. People have a common goal and people have to work on the same problem, and they have to share their information to solve the problem, figure out what's going on and deal with the hazard. There's a real compelling reason to work together. And that's different if you're just mapping Holocene tephras from St. Helens, you and your buddy, because you're only looking at one aspect, and you've got the luxury of time, and you don't have to worry about seismic activity or deformation or gas geochemistry. So you can focus on whatever it is you're doing. But when the crunch is on, and you're having to do things on the fly, in real time, you have to work as a team. It didn't work in St. Helens in 1980, but it developed."

CHAPTER 7

Mammoth Lakes: Between a Rock and a Hard Place

With St. Helens becoming predictable, Dan Miller left for California to resume his work of uncovering the history and hazards of volcanoes there. He had largely finished with Mount Shasta in the northern part of the state and focused this time on the Mammoth Lakes/Long Valley region, a sprawling and complex volcano system on the eastern edge of the Sierra Nevada.

Unlike St. Helens, these volcanoes were not dramatic mountains. The "volcano" in which the town of Mammoth Lakes is located is actually an oblong-shaped valley spread out at the eastern base of the Sierras. About 760,000 years ago, the volcano had let loose with a gigantic explosive eruption, two thousand times larger than the St. Helens's May 18 blast. Only five other eruptions in the last 2.5 million years have been as large.

"This was a humongous eruption," said Survey geologist Patrick Muffler. "It makes anything we've seen in later years look like child's play. . . . It was, believe me, an incomprehensibly catastrophic event."

The release of so much material had partially drained the magma chamber. Eventually the earth above collapsed into the chamber, creating a gigantic cavity on the surface known generally to geologists as a caldera, and known locally as Long Valley. Long Valley is shaped like a racetrack with its western bend enclosing one of the region's most popular resorts.

As in most volcanic regions, the geologic machine had at Long Valley created a rugged beauty that had long invited tourists to go hiking, fishing, and skiing. In the early eighties, they were doing so in growing numbers. The Mammoth Lakes/Long Valley region had become a playground for vacationers from Los Angeles, which is only a five-hour drive away. When Miller returned from Washington State, the region was becoming quite prosperous. New homes were selling for $175,000 and shopping centers had sprouted at crossroads. Sales were high in the outdoor-equipment shops, and builders worked overtime to keep pace with the demand for vacation condos.[1]

In May of 1982, Miller attended a scientific conference organized by the U.S. Department of Energy and the Survey. The meeting, which was held at the Sierra Nevada Inn, drew dozens of federal, state, and academic geologists. The DOE had long been interested in the geothermal potential of the region, and the meeting was designed to draw some conclusions about the direction for geothermal development.

During the meeting, a seismologist from the University of Nevada at Reno named Alan Ryall shocked many in the conference room, especially Dan Miller and the other St. Helens veterans, with his presentation. Ryall had spent years recording, cataloging, and interpreting the quakes that bang under the string of mountains that stand along the California-Nevada border.

Ryall began by pointing out the entire eastern side of the Sierra Nevada had been jumping with quakes for decades. Then, in the early 1970s, the region became unusually quiet. Behavior changed again in 1978 with a large quake, registering a magnitude of 5.4, which was followed by hundreds of smaller ones. In 1980, starting on May 15, a series of intense earthquake swarms shuddered through the region. On May 25, exactly a week after St. Helens's first big eruption and just hours after it had launched into its second eruption, four magnitude 6.0 quakes jolted the Mammoth Lakes region within forty-eight hours.

One of the things that makes this region a challenge to volcanologists is that it has a lot of regular, everyday tectonic quakes. But the May 25, 1980, quakes were unprecedented. During one forty-eight-hour period, cupboards swung open and dishes slipped out, cans on grocery store shelves bounced and fell, building foundations and walls cracked, avalanches and rockfalls tumbled down local mountains, and fissures opened. The area was jarred not only by four quakes of 6 or greater, but by tens of 4s and 5s, and hundreds of smaller quakes. By the time the quakes stopped, they had caused $2 million in damages. Nine people had been injured by the quakes and one woman had suffered a miscarriage.[2]

At the time, seismologists speculated that the quakes were tectonic, originating along the fault zone that extends along the eastern foothills of the Sierra Nevada. But Ryall, who had been plotting the point of origin of quakes for decades, saw three interesting patterns emerging from his data. First, the quakes were migrating from six miles south of the Long Valley caldera into the southern crater moat. The earthquakes themselves were also getting shallower, rising from a depth of 4.8 miles to about 3 miles from the surface. And most troubling, the quakes had a signature that was consistent with magma moving beneath the surface.

When Ryall finished his presentation, the St. Helens vets were trading glances around the room. Each had seen its significance. It was possible, even likely, that a tongue of magma was moving through an ancient underground channel, working its way to the surface. If it was within three miles of the surface, it had entered the magic zone of unpredictability.

The geologists learned something else during the meeting. Following the 1980 quakes, a large dome-shaped area inside the western rim of the caldera had begun to bulge. In fact, an island of land that rose fifteen hundred feet above the valley floor had risen another ten inches, probably, between 1979 and 1980.[3] This uplift, an area that volcanologists call a *resurgent* dome (not a *lava* dome like those on St. Helens's side), was directly above the ancient magma chamber. The resurgent dome's growth rate, by St. Helens standards, seemed puny, but new deformation along with seismic data suggested new magma was now below the resurgent dome and perhaps pushing its way up through ancient channels. "That wasn't anticipated at all," said Roy Bailey, who had recently been named to head the Survey's volcano hazard programs and was the Survey's leading expert on the region.

During the dinner that closed the conference, the table talk was all about the possibility of an eruption, when it might occur, and what consequences it would have. Knowing that the system had produced one of the biggest eruptions ever made everyone uneasy. Dan Miller says, "We were concerned about the possibility of rising magma. If it reached a depth of a mile and half, it could hit the water table, leading to steam explosions and an eruption."[4]

Dan Miller knew the area well enough to visualize what an eruption from the southwestern part of the caldera would mean. Even an eruption the size of a St. Helens could be a killer. The worst possible case involved an explosive eruption in the winter. If that happened, it could generate floods and mudflows.

An explosive eruption from the resurgent dome could threaten everyone uphill in the resort because the single road down the mountain and out to safety cut directly through Long Valley and almost right over the earthquake zone. If an explosive eruption started under those conditions, thousands of people could find themselves trapped between impassable mountains and an erupting volcano.

During this final dinner, with everyone discussing the possibilities, the Sierra Nevada Inn and the rest of the region were bounced by another quake. It was a sudden bump. Everyone watched the lamps sway and heard the ice in their glasses clink.

Ryall left the dinner and called his office to check his instruments. He returned and pulled Roy Bailey aside. The quake had been a 4.2 with several others that couldn't be felt, Ryall said. Then he said, "These earthquakes are now within two kilometers [1.2 miles] of the surface."

Bailey quietly told the St. Helens vets that they should gather later at the Forest Service headquarters. "We were looking at the precursors of a possible eruption," says Bailey. "We were all pretty excited."

At the meeting, Miller and others had to face that, unlike with St. Helens, this volcano's history was not well understood. They had little information about periodicity, if any existed, which might help them predict a future eruption. The most immediate concern was the real possibility that heat from the rising magma would turn underground water into steam, which could produce a dangerous eruption soon. The geologists decided that the Survey should issue a warning and issue it quickly. The meeting broke up after midnight, but few people got much sleep. Earthquakes continued to bounce the Sierra Nevada Inn through the night.

The business of issuing warnings about the future of the neighborhood volcano was still quite new to the Survey. Certainly by this time, the warnings at St. Helens were approaching a science. Because those warnings were built on solid patterns of detectable behavior, and eruptions nicely followed warnings, there was a growing sense of confidence in the Vancouver area in the Survey's new skill. But that warning system had not come easily at St. Helens. In fact, next to arguing about the danger of working in the crater, geologists in Vancouver spent weeks wrangling over when warnings should be issued and, line by line, what they should say.

The Survey was bound by law to warn the public of natural hazards.[5] But Mammoth Lakes was not St. Helens. There was no local experience with the dangers of a volcano, so disbelief would be high. There was little data about the volcano's past behavior on which to base predictions, so forecasts would be open to criticism. And as the Survey had learned at Mount Baker and as the French geologists had learned at Soufrière on the island of Guadeloupe, a false warning could do major damage as well. Yet the earthquake and deformation data told the geologists that this volcano, which was capable of extraordinary blasts, was progressing toward another eruption. In sum, when it came to issuing volcano warnings, the Survey's limited but growing expertise with explosive volcanoes, combined with the unpredictability of those volcanoes, put the scientists between a rock and a hard place.

After the meeting, Dan Miller returned to Denver, where he, Rick Hoblitt, Rocky Crandell, and Don Mullineaux began writing the warning document. For an outline, they used the old St. Helens Blue Book, which had provided the geologic history, the current evidence for concern, and a list of the potential hazards. Roy Bailey, who had spent much of his career studying the volcanic system under Long Valley, became a key participant in preparing the report.

The work was done with some sense of urgency. "We were concerned that the activities were escalating and might actually lead to an eruption before we could get it out," said Miller. The document was ready within a week and sent out for review by other Survey geologists. Internal reviews in the Survey are like peer review for publications in most scientific journals. It can be a nit-picking, hypercritical process, but once a paper passes through these fires, it is certified as scientifically sound and the seal of the Survey is placed on it.

The process commonly takes months, but the Mammoth warning was put on a fast track. Toward the end of May, the hazard warning was nearly finished, but it was not fast enough. Before it could be published, an enterprising and dogged science reporter for the *Los Angeles Times*, George Alexander, sniffed out the story and on May 24 published that Survey geologists were concerned that Long Valley was showing ominous signs common to volcanoes preparing to explode. The story was picked up by the wire services, and within hours it was national news. The news account triggered a series of reactions that resulted in a social and economic convulsion around Mammoth Lakes.

Almost immediately, the town was inundated with calls from around the country. When reporters or businessmen contacted local politicians, the elected officials couldn't provide information because they had no notification yet from the Survey. In fact, the first word any elected official got about new dangers from the local volcano came from Alexander's story in the *Times*.

"It's crazy, it's ridiculous," Julie Fitzpatrick, manager of the Mammoth Lakes Chamber of Commerce, told a UPI reporter shortly after the *Los Angeles Times* story broke. "We're swamped with phone calls, hundreds in fact. It's really irritating. . . . People were calling in to know if they could get a good spot to watch the lava flowing and volcano going off. Other people were calling us telling us, 'Get out of town, you're going to get killed.' Everybody in town around here is fine."[6]

The day after the story broke, the Survey issued its "Notice of Potential Volcanic Hazard." The warning noted that a "notice" was the lowest of the three levels of alert the Survey could issue. The Survey said that it was intensifying monitoring, but "available evidence is insufficient to suggest that a hazardous event is imminent."

The notice outlined the evidence that had concerned the scientists. The region had been widely studied because of its volcanic past. In 1976, scientists had identified a large and partially molten magma chamber about five miles beneath the Long Valley caldera. The recent earthquakes were of a type that had not been seen before in the region, but they were similar to those elsewhere that had been associated with moving magma. Also, the depth of the quakes had become "progressively shallower since June 1980." In October of 1980, surveys along Route 395, the only road out of Mammoth, indicated the ground had been raised nearly a foot in little more than a year. And in January 1982, new steam vents had opened less than two miles from the resurgent dome. Taken together, the notice read, the most likely scenario was that a tongue of magma was "moving toward the surface. It cannot yet be said for certain that this magma will cause an eruption."

The report detailed the kinds of hazards an eruption would pose to the resort. Large rocks could rain as far as six miles from an eruption vent. Lightning could ignite fires tens of miles away. Pyroclastic flows passing over snow could release hot mudflows. Hot rocks and gases could melt snow and launch floods down valleys. Gases could suffocate animals and humans.

The geologists were delighted to have finished their work so quickly. But oddly, none of the scientists thought that it would be a problem to publish their warning—that a volcano could erupt under a popular resort—just before the Memorial Day weekend. And that very weekend, many visitors, who had been amused by the warning and kept their plans to spend the weekend in the Mammoth Lakes area, became spooked when four moderate earthquakes rippled across the region. A substantial number packed their cars and fled.

When those people left town, it seemed they took the good times with them. Business plunged. The locals knew the reason—the hyperactive federal scientists and their crazy damn volcano warning. The scientists themselves seemed oblivious to the chaos they were creating. One of them told a wire service reporter, "If highway 395 breaks, there is no way to get out. In terms of an economic and social disaster, it could be much more devastating than Mount St. Helens."[7] The media exacerbated the situation. A Los Angeles television station broadcast a report juxtaposing film of skiers at Mammoth Mountain with scenes of St. Helens erupting.

Dan Miller, who was made chief of the Mammoth Lakes/Long Valley project, and the other Survey geologists were not warmly welcomed when they descended on the area for fieldwork that summer.

As the summer dragged on, it was obvious that the life had been drained from the economy. The real estate market suffered the most dramatic losses, although overbuilding also contributed to developers' problems.[8] Architects reported that plans to construct vacation homes were being canceled or put on hold. Local architect Allan O'Connor complained, "Immediately after reports of the volcano in the newspapers, I had seven jobs called off in a five-day period."[9] Condo sales began drying up. With the falloff, businesses began closing. Commercial vacancy rates began pushing 30 percent. Some businesses were soon reporting that gross revenues were down 25 percent. Taxable sales dropped, while they grew overall for the rest of the state.[10]

Miller and the others tried to organize scientific meetings for the community and the local and state officials to explain the reasons for the warning and to answer questions. The meetings were delayed. First, the town fathers didn't want a volcano meeting during their summer festival days, which fizzled anyway. Another meeting was torpedoed because it was to be held in a nearby town. Then, state officials had scheduling conflicts.

Meanwhile, the geologists took every chance that was offered to meet with residents and businesspeople. They thought that if they had a chance to explain the migration of earthquakes, with their unique volcanic signatures, and the ten-inch bulge, everyone would see the light, the fuss would end, and disaster planning would begin.

Completely wrong. Many local officials refused to meet with Miller and the others. Those who did were often insulting. "You're a terrible scientist" and "You did this only to pump up your budget" were common opening remarks from the townsfolk. Call-in radio shows were largely used to berate the geologists, to attack their motives, and to quarrel with their warning. After all, no one really knew when, or even if, a volcano would erupt. Signs reading GEOLOGISTS NOT WELCOME began appearing in motel and restaurant windows.[11] The public gatherings were often disastrous. After an introduction at a local Lions Club meeting, for example, Miller was loudly booed before he could begin his slide presentation.

Once people stopped verbally attacking the scientists, everyone wanted to know about timing. If this thing was going to blow, when would it happen?

The trouble was that the region underground was geologically complex and still largely a mystery. Two large magma chambers under the region nurtured volcanism over a wide area, but little else was known. Said Roy Bailey, "It's like trying to describe the behavior of a newly discovered animal. You don't yet know what exactly it takes to make the beast angry."[12]

In July, the county board of supervisors unanimously passed a resolution ordering the "Survey cease and desist from issuing inflammatory press releases on alleged potential hazards in the Long Valley Area."[13]

One afternoon late in the summer, after an especially difficult day,

Dan Miller stopped into a local restaurant called Whiskey Creek. No sooner had he ordered than a local realtor came up to him and said, "You better watch what you're saying or someone is likely to put a bomb in your car." The mood in town was so bitter that Miller took the threat seriously. He returned to the trailer where he was living with his wife and two daughters, concerned about their safety. Miller was relieved when that summer field season ended and he could return to Denver.

On January 6, 1983, he got a call from the Survey regional headquarters in Menlo Park. Long Valley and Mammoth Lakes were rocking. Two 5.3 quakes had slammed the region. The quakes destroyed an airplane hangar. Power and telephone communications were out over a wide area. Following the big tremors, swarms of quakes began blurring the seismographs at a rate of forty a minute.[14] It was impossible to tell from the instruments if the energy level was increasing, nor could anyone tell if the quakes were tectonic or magmatic, but there were hundreds of them.

Miller and Don Mullineaux flew to Menlo Park, where they conferred with senior members of the Survey. In Menlo Park, they were told the seismographs were recording spasmodic tremors, magma's hoofbeats, and the quakes were less than two miles from the surface. They were practically underneath the only road out of Mammoth. Miller and Mullineaux were put aboard a charter plane and flown directly to Mammoth Lakes. They set up shop in a fire station and began conferring with local officials.

As happened at St. Helens, non-Survey geologists began making independent predictions about the likelihood of an eruption. One of them, a geologist from Sandia National Lab, said an eruption was close.[15] The scientist was basing his prediction on the seismic data. By his reading, the origin of quakes was shallower than the two-mile depth that was the Survey's estimate. Actually, both groups were guessing because, as at St. Helens, scientists didn't have a good picture of what the earth's structure looked like under Long Valley, from the surface to about three miles deep, and therefore how it might reflect seismic signals.

The earthquakes had grabbed everyone's attention, and a few people even remembered the Survey's warnings of the special dangers of a winter eruption. On an average day, the mountain was covered with fifteen thousand skiers, with more on the weekends. Within a few weeks, the quakes had died down, but a new survey of the broad resurgent dome found that it had risen another three inches over the winter.[16]

The earthquakes and the subsequent briefings by Miller and Mullineaux to the county officials and the public began to change the minds of a couple of local politicians. County supervisors Michael Jencks and Al Leydecker had even gone out with the geologists on field trips so they could see the area as the geologists did.[17] The two requested public updates from the Survey on the evolving unrest. While the two politicians did not accept the Survey's warning wholeheartedly, they thought that it would be prudent for the county to create an alternative route out of the mountains just in case, someday, the road over Long Valley was impassable.

The two county supervisors worked hard that spring to find the money to pave a six-mile, little-used logging road through the mountains to the north of Mammoth Lakes. It was completed by the following October. A new sign proclaimed the road the Mammoth Scenic Loop, but it became known locally as the Mammoth Escape Route. But the route was completed months after the winter scare, and those concerns had been replaced by more business concerns. Businesspeople in the community were outraged that the two supervisors would do something that seemed to reaffirm the possibility that their heavenly valley could explode. A recall election was organized. The two men, who had long been mavericks in local government, fought what became a bitter race and lost.

Mammoth continued to sink deeper into tough times. The Bank of Mammoth, which handled the finances of two-thirds of the town, began bleeding money as deposits decreased and savings withdrawals increased. In 1983, the bank recorded a half-million-dollar loss due primarily to loan delinquencies. To keep afloat, its name was changed and stock was offered to the public.[18]

Community leaders began an intense lobbying campaign aimed at influencing members of the Reagan administration. One of the people rumored to be involved in the lobbying effort was a local electrical engineer named Bill Graham, who became Reagan's science adviser.[19] Senator Richard Lugar, a Republican from Indiana, was also instrumental in carrying messages from constituents to Interior Secretary James Watt. Gary Flynn, chairman of the Mono County Republican Party and Mammoth's pro tem mayor, acknowledged in 1985 that he had helped with the campaign. "We made phone calls and wrote letters. We did every-thing we could to get responsible statements from the Survey," he said.[20]

Another one pressing the Reagan White House was Dr. Martin Dubin, who brandished his political connections (a member of the big contributors "Inner Circle," etc.) in his letter to Interior Secretary James Watt. In that letter, he said the Survey was employing "scare tactics . . . for selfish reasons to obtain their budget." He demanded Watt force the Survey to "make an announcement to the public relating the true facts."[21]

Flynn and others received a warm reception. Reagan was a former California governor and had a sensitive political network in the state still functioning. In fact, one of the members of Reagan's transition team, Bill Graham, lived in Mammoth. Perhaps the most accommodat-ing member of the Reagan cabinet was Secretary Watt. The U.S. Geo-logical Survey was an Interior Department agency, and both the director of the Survey and Watt himself were heavily lobbied for relief from the "absurd" volcano warning that was destroying the resort's eco-nomic life.

Then, in September 1983, a peculiar thing happened. Suddenly, and without consultation with the geologists in the field, the director of the Survey discarded the agency's three-tiered warning system. In its place, he declared that warnings between scientists and public officials would be informal and that there would be no public declarations. The move, in essence, removed the official notice of hazard from Mammoth. The director of the Survey said the warning system had been reexamined and found wanting. Most Survey scientists would have agreed that the warning system needed overhaul, but to abandon it without consulting the geologists most directly affected by it, those working in Long Valley, came as a shock.

Survey geologists in Mammoth Lakes were furious. To them, the director had caved in to political pressure. So for years to come, they simply ignored the order.

After a while, though, they began to question whether there might be a warning system that could be more helpful, one that would trigger a minimal level of concern when warranted, and one that could be used by emergency management officials to more effectively plan for a natu-ral disaster. Eventually, they devised a four-tier system, with objective criteria and specific responses for each level. And after James Watt resigned, the Survey quietly adopted it.

By 1983, the Long Valley caldera had been rumbling right along but didn't look any closer to an eruption than it had a year or even two years earlier. At a Survey meeting in the Napa Valley, Chris Newhall suggested that someone should take a look at similar complex systems elsewhere. Perhaps they might share some characteristics with Long Valley that would be helpful. Newhall and his colleague at St. Helens, volcanologist Dan Dzurisin, were tapped to do the job.

Newhall had no idea how difficult the job would be. He set aside three months in 1983 to do it, but the two scientists didn't finish their work for another four years. Then they published a massive eleven-hundred-page report describing unrest at the world's calderas. It was a monumental piece of work.

"What we discovered was that almost every large geologically young caldera is restless," says Newhall. "And that the types of things that get observed in Long Valley are actually very common at other calderas. That doesn't say it's not going to erupt. All it says is that you need to temper your concern about an eruption by realizing that these are big volcanic systems and that they have a lot of buffering capacity for disturbances. They are easily disturbed but it's harder to get them actually to erupt. And the result is that you probably get something in the order of nine episodes of unrest for every one that actually culminates in an eruption."

The Mammoth Lakes crisis had put Dan Miller and the other Survey geologists in a difficult position. St. Helens had taught them the power of an explosive volcano, and then it showed them how to better monitor a volcano's vital signs to predict an eruption. But what was becoming apparent was that each volcanic system, or possibly even each volcano, had its own metabolism, and scientists had to approach each one as a unique beast. Just as transferring expertise from Hawaii to St. Helens had its shortcomings, transferring St. Helens expertise to Long Valley had misled the scientists. St. Helens and Mammoth Lakes were different animals. The scientists were learning more about the precursors of an eruption, but they had a lot yet to learn because volcanoes come in wide varieties.

From the Mammoth Lakes unrest and subsequent studies that came out of the experience, the geologists learned a lot about calderas. The scientists were also able to construct a more effective warning system. And perhaps most important, they learned that if their warnings are to be effective, they have to have community leaders on their side from the beginning.

By 1985, even University of Nevada seismologist Ryall, whose career had been spent monitoring Sierra quakes, was saying the Survey had "overreacted" because "they had been burned—figuratively and literally—at St. Helens."

Responding to Ryall's criticism, Patrick Muffler, who was then the head of the Survey's volcano hazards program, said, "We were damned if we did and damned if we didn't. But what if something had happened and we hadn't said anything? If the 1982 events had led to an eruption, we'd have been canonized instead of castigated. Our job is to let people know when a hazard exists. Yet the minute we mention hazard, we are alarmists."[22]

Eventually, the Survey would establish a state-of-the-art, remotely operated observatory to monitor activity in the Long Valley/Mammoth Lakes region. The first director of the observatory was Dave Hill. In analyzing all the problems around the warning, Hill wrote:

"The threat of an impending volcanic eruption inevitably presents a narrow path to a successful response. On one side, an overanxious response will exacerbate the 'false alarm' problem (one, or at most two, 'needless' evacuations, for example, may well destroy our credibility and effectiveness). On the other side, an overly laid-back response (no need to worry yet) may lead to serious casualties and fatalities if an explosive eruption does develop. . . . This places a premium on good science, experience 'under fire,' cool heads, and yes, a little luck."[23]

At that point, however, the overwhelming majority of the Survey's experience under fire from explosive volcanoes was with just one, St. Helens, and even St. Helens was running out of steam.

CHAPTER 8

The Volcano Zoo

If Survey geologists were to get more experience with explosive volcanoes, they had two choices. They could either wait for some other
Cascade volcano to erupt or go looking for restless mountains. After the
St. Helens eruption, there was a modest amount of extra funding and
official support for working overseas.

A year before St. Helens exploded, one of the young part-time geologists working for the Survey in Hawaii, Dan Dzurisin, had started to
develop a program in Indonesia, the island state in the southwest
Pacific. Indonesia is one of the most volcanic areas on earth with 126
active volcanoes. In 1981, Survey officials revived Dzurisin's plan and
began dispatching volcano specialists. Officially, the international program was to help Indonesia establish monitoring and hazard-
assessment programs. Unofficially, the Survey would learn just as much
as it taught.

Hoblitt got his turn in 1982. On his way, he first stopped to inspect
Lamington, a volcano in what is now Papua New Guinea (PNG). A
paper written by Australian geologist G. A. M. Taylor, about the 1951
eruption there, may have been the most precise observation of an eruption written to that point.[1] In a two-year investigation, Taylor did the
work of an entire volcano observatory, and for it England bestowed on
him the George Medal, the civilian equivalent of the Victoria Cross.

Taylor's report was a classic among volcanologists, and long after it had gone out of print, copies were regarded as collector's items. Hoblitt had read all 116 pages several times, analyzing the data and inspecting the photographs.

Not surprisingly, the people who had lived on and near the mountain had no idea that it was a volcano. Anthropologists had never heard folktales about Lamington's having exploded prior to its eruption this century. But, within a week of its first activity, Lamington had its cataclysmic eruption, one that had killed nearly three thousand people.

On Monday, January 15, 1951, several people at a plantation nearby had noticed landslides that appeared as brown streaks on the steep, forested slopes of Lamington. At 3 P.M. that day, smoke was seen rising from the mountain. Tuesday, the earthquakes began, some of which were described as seeming to make "the whole earth rock." Wednesday, a policeman recorded that quakes were coming every seven minutes. Thursday, a party of police went to investigate the mountain, but they returned early because the ground was shaking so violently. Thursday night, "the whole sky was alive with electrical activity." Friday, people reported seeing "whirling stars" (perhaps hot ash) in the sky, and close to the mountain it began to rain stones, causing some houses to collapse. Saturday, a roar from the crater like a "gigantic underground railway" could be heard ten miles away, and "wireless reception became a buzz of static." The quakes had become so powerful that some people were tying themselves to trees to keep from being thrown about. But at 8 P.M., the earthquakes suddenly stopped.

At 10:40 A.M. on Sunday, January 21, the pilot of a Qantas airliner flying from Port Moresby bound for Rabaul saw an explosion burst from Lamington. He was flying at ninety-five hundred feet and watched as the dark mass shot out of a crater to forty thousand feet. Most important, to Hoblitt anyway, the Qantas captain reported that the base of the column expanded rapidly, as if the "whole countryside were erupting." This was a massive ground-hugging ash hurricane with temperatures reaching four hundred degrees Fahrenheit. People on the ground said that it was "swirling and billowing like an oil fire," but it advanced almost as a solid mass, consuming sixty-eight square miles around the mountain and killing nearly every living thing it touched.

When it was over, 2,942 people were dead, almost all of them from the effects of the pyroclastic flow. Taylor noted the bodies were so badly burned that "it was extremely difficult to distinguish between Europeans and natives forty-eight hours after the eruption." Many were killed on roads as they tried to run from the mountain. At a mission school, dozens of children had taken refuge in the schoolhouse, but the pyroclastic flow killed them, leaving their bodies in a pile, and then blew away the walls of the building. One native in a group of four survived when the flow knocked him into a stream. He rose out of the water, but breathing burned his throat so he plunged back under, waited as long as he could, then took a breath of hot dust. Luckily, the superheated air had passed.

When Hoblitt arrived in PNG, he was met by a French volcanologist named Patrice de St. Ours, who had been working in the region. The two drove to a remote corner of the country and hired natives who once lived in what had been the volcano's blast zone to guide them. The area had been incinerated thirty years earlier, yet by the time Hoblitt arrived, he confronted a dense jungle. The four men hacked their way in, with an occasional stop to allow Hoblitt to unload the leeches from inside his hiking boots.

Here and there, he would dig pits and study the deposits, and eventually he found remnants of the blast extending out of an opening in the crater, as he expected. He also found the same blast material all the way around the volcano. But the debris far from the volcano should not have been spread so uniformly. Pre-eruption Lamington was shaped like the posteruption St. Helens. It had a breached crater, a U-shaped crater with high walls on three sides.

After studying the Lamington deposits, Hoblitt started to believe that something unusual had happened. The conventional view was that the Lamington eruption was a "directed blast," like St. Helens, and Gorshkov, who had speculated on such things at Bezymianny, had put Lamington in the same category. But unlike St. Helens, where the blast products were primarily in one sector, Lamington's deposits were distributed radially, all around the volcano.

Looking deeper into the rock and ash records around Lamington, Hoblitt began to understand what had actually happened, and with that insight he learned that less vigorous eruptions can be more dangerous than more powerful eruptions.

Hoblitt's view was built upon the "column collapse theory," which was developed by Steve Sparks and Lionel Wilson in 1976. In the standard view of eruptions, the material (gas, ash, and rock) in all volcanic columns rises by force of the initial explosive jet. But shortly after the eruption, some of the heavier material in the rising column is overcome by gravity, like a bullet shot into the sky. The material slows until its momentum is equal to the pull of gravity.

At this point, one of two things can happen. If the heated air in the mixture is sufficient to lower the density of the mixture below that of the surrounding atmosphere, the plume will become buoyant and rise convectively, like smoke from a campfire. If not, the jetted mixture produces a fountain of ash and gas that falls back to earth and splashes around the volcano at high speed like a wave crashing against rock. This is what the Qantas pilot had seen when he said it looked as if the "whole countryside were erupting." The material falling back to earth crashed with such force that it climbed the high walls of the crater and began flowing downhill, incinerating homes, cattle, and people. It was an unusual feature of an explosive eruption known as a pyroclastic surge, a mechanism seen in an atmospheric atom-bomb test at Bikini atoll.

The key point was that these crashing waves of superheated ash and gas were more likely to result from a weaker eruption than from a more muscular one. An energetic jet with a high content of magmatic gases will usually incorporate enough air to become buoyant. But a weaker jet, with less magmatic gas, will likely result in a collapsed column and become a fountain of incandescent pumice.

Lamington was only the start of this phase of Hoblitt's education. It was an adventure, and volcanoes were only one aspect of it. He camped at a highland village one night and had his wallet stolen. The next day, the natives were so humiliated by having a guest robbed that when they found a likely suspect, they shot him in the ass. Hoblitt met plantation managers in bars that seemed like props in Bogart movies, complete with ceiling fans that rotated about once a minute. One night, he saw natives doing a fire dance, including sitting on hot coals. "These guys were not wearing their Fruit Of The Looms. They were bare pickle," he said, laughing later.

He flew to Indonesia, drove to a city of two million named Bandung, in Java, and slept the sleep of an exhausted Third World traveler. The next morning, he thought he was suffering from jet lag because when he walked outside of his hotel, it was still dark. Then he noticed people running with umbrellas. When he was fully awake, he saw that the local volcano, Galunggung, 120 miles away, was in full eruption and it was raining ash.

Later, Hoblitt visited the Galunggung Volcano Observatory. When he arrived, the volcano had quieted, but it was obviously still very active. Hoblitt looked at the seismograph. "It was the most beautiful harmonic tremor I've ever seen. It looked like they were using a sine wave generator," he remembers. The volcano was close to another eruption.

Hoblitt prodded his hosts into a car and began driving toward the volcano, past homes that had collapsed roofs and holes punched through by flying rock. When he was as close as the road would take him, he got out of the car and heard a whooshing noise. Then another, but this time the sound was louder. Then louder still. Mount Galunggung was chugging like a train leaving the station. The Indonesian geologists had to practically drag him into the car, and moments later, Galunggung was in full eruption again, this time with lapilli (pea-size stones that seconds before were molten rock) pinging like rain on the car. Nearby, he saw a child in a field flying a kite. What struck Hoblitt about the scene was that the child was mesmerized by the kite. Hoblitt realized that eruptions, which were so exciting to an international volcanologist, were already old hat to Galunggung's children.

Galunggung was just the first volcano Hoblitt saw in Indonesia. The nation really was a volcano zoo. In fact, Indonesia grew out of the collision of three great plates. As a result, it has more active volcanoes mixed with a denser population than anywhere else on earth.[2] Indonesia has hundreds of volcanoes. Seventy-five are designated as Class A (active since 1600). They have produced more than eight hundred eruptions since 1600, and seventy-four of those have been major calamities.[3]

Indonesia's volcanoes may be the most deadly on earth. Volcanoes produce rich soil, and in a society with widespread poverty, the poor will use every acre of land to farm. In Indonesia, that meant marching right up the slopes of active volcanoes and hoping for the best. In the last two centuries, an estimated 140,000 people have died in eruptions in Indonesia.[4]

With large populations living close to active volcanoes, Indonesia holds the record for the number of deaths directly caused by volcanoes. Indonesia's Krakatau heads the list of killers. In 1883, the volcano exploded with such force that the sound was heard as far away as Istanbul.[5] News of the massive eruption was flashed around the world over the new undersea telegraph cable. One of the first messages read: "Where once Mount Krakatau stood, the sea now plays." So much material was emptied from the volcano's magma chamber that the overlying rock collapsed into a caldera five miles long. When the vertical eruption column reached the sea, it set off a tidal wave fifty feet high. It washed over the lowlands of Java and Sumatra, carrying most of those killed in the eruption, thirty-six thousand people, out to sea.

The link between a volcanic eruption and global climate was first established with the Krakatau eruption. For the first time, scientists worldwide could link regional meteorological observations to a distant event. In addition to air-pressure changes, the enormous quantity of ash was observed to have influenced the atmosphere on the other side of the earth. Brilliant sunsets were recorded everywhere. They were so intense in Poughkeepsie, New York, and New Haven, Connecticut, for example, that fire departments were called out at sunset when townsfolk thought their cities were ablaze. Scientists also confirmed that global temperatures dropped an average of one degree Fahrenheit for several years after the eruption.

And yet Krakatau wasn't the largest eruption in human history. Just sixty-eight years earlier, Indonesia's Tambora volcano, east of Java, exploded with a force that may have been five or ten times greater. The number of people killed directly by the blast is unknown, but it was certainly in the tens of thousands. And the subsequent crop failures and famine, a direct result of the eruption's effect on climate, are estimated to have caused eighty thousand more deaths.[6]

For centuries, it was thought that volcanoes affected climate as a result of the fine particles they ejected into the atmosphere. But it was another Indonesian volcano that showed the true mechanism, because even fine particles of volcanic debris drift back to earth relatively quickly after an eruption. In the 1963 eruption of Gunung Agung, on Bali, scientists found it wasn't dust but sulfur dioxide that produced long-term changes in the atmosphere. When this gas reached the stratosphere, it remained there for a long time as an aerosol of fine droplets of sulfuric acid. It scattered incoming solar radiation and thereby warmed the stratosphere. And it absorbed infrared radiation, thus cooling the lower portion of earth's atmosphere.[7]

"I was there primarily not to learn but to try to share what I knew with these people," Hoblitt recalled. "But I was seeing a whole hell of a lot of volcanoes. You couldn't help but learn stuff."

At Merapi, on the Indonesian island of Java, Hoblitt saw a volcano in its dome-building phase, like the current state of St. Helens. Again people were living right up its flanks. Merapi had a history of terrible eruptions, and folklore had it that one eruption was so horrific the Hindu rajah moved to Bali and Java became Muslim.[8] By the time Hoblitt arrived, Merapi had erupted forty-four times in the past eighty years. A high and steep volcano, it lies in densely populated central Java. In modern times it has not been overly explosive, but its dome-building eruptions can be treacherous. When the dome, which is growing right at the summit, oversteepens, a piece of the dome can break off. If the falling slab is large enough, it can generate ash hurricanes and channel hot avalanches that rush down the steep slopes and across farmlands. These lethal dome-building events are known in the trade as Merapi-type eruptions. In 1930, one such event wiped out thirteen villages and killed 1,369 people. The Indonesian Volcanic Survey would establish a seismic net here that was able to forecast Merapi somewhat.

Another new and strange volcano Hoblitt studied was Kelut, which was infamous for its lahars. The word *lahar*, which is now the international term for "volcanic mudflow," is of Javanese origin. Lahars are the most common hazard of an eruption in Indonesia. The lower slopes and alluvial plains surrounding volcanoes in Indonesia are generally lahar deposits. Kelut, a little over one hundred miles east of Merapi, is one of the lowest volcanoes in Indonesia (5,600 feet), and its eruptions are brief. Yet both of these features made Kelut dangerous. Its fertile ground entices people to work rice fields around Kelut. And its brief eruptive periods mean that when it does go, it can be especially violent. These brief, violent eruptions happen once every seventeen years on average.

Another of Kelut's features that interested Hoblitt was its nasty crater lake. Sulfur dioxide is almost constantly seeping into the lake, creating a body of sulfuric acid. Worse, eruptions generally blow right through the lake, launching mudflows of new rock, old debris, and sulfuric acid. In 1919, Kelut killed more than 5,100 people and destroyed 135 square kilometers of farmland. The disaster led to the creation of the Netherlands East-Indies Volcanological Survey, which later became the Volcanological Survey of Indonesia. It also led to efforts to mitigate the risk from the lake. Massive tunnels, for example, have been drilled to siphon off the lake's waters. These efforts were successful in reducing the water level and subsequent floods, but they were often upended by the next eruption. In its most recent eruption, in 1966, Kelut generated a massive lahar that killed 282 people, damaged fifty-two villages, swept away ten bridges, and obliterated thirty miles of highway.[9]

In central Java, Hoblitt inspected Dieng. This volcano had a bad reputation even by Indonesian standards. The volcano's piping system was spread over a vast region pocked with shallow craters. People in the region were not only threatened by the usual calamities a volcano can launch but also by a unique threat from volcanic gases. As magma rises to the surface, pressures begin to ease and gases that have been trapped in the magma begin to separate from it and seep out onto the surface even without lava appearing. Some of the gases, such as CO_2, are heavier than air and can cling to the ground and even flow downhill or into low-lying areas. In February 1979, a bubble of CO_2 did just that and suffocated 149 people.[10]

One feature that Hoblitt saw again and again was the hummocky fields that Glicken was studying at St. Helens. Sometimes, on older volcanoes, Hoblitt would see more than one set of hummocks, but they would be at the base of a perfectly symmetrical volcanic cone. The conclusion was inescapable: volcanoes fall apart and rebuild themselves constantly. St. Helens was at the beginning of that cycle now, but a thousand years from now it might stand as a perfect symmetrical mountain, capped once more with a white peak and with some strange but weather-worn hummocks spread out quietly at its base.

The Galunggung volcano had a feature widely spread around one flank known as the "ten thousand hills of Galunggung." Galunggung, like St. Helens, had a breached crater, and the collapsed summit and wall was the source of the ten thousand hills, just like the gigantic debris avalanche at St. Helens. But what struck Hoblitt was that population pressures in Indonesia had driven people up the sides of the volcano. Some had even been living and farming right at the crater breach. An eruption had recently flattened one village there. When he inspected it, Hoblitt found pieces of terra-cotta and other artifacts of village life. He began digging his holes and found only thin deposits of debris. This was a puzzle since the area had obviously been subjected to a powerful blast.

"There was no visible sign of the village, but it was clear we had the right deposit because it had terra-cotta tile fragments from the roofs of the houses that had been destroyed," said Hoblitt. What Hoblitt and other geologists were just beginning to understand is that powerful effects don't always leave thick layers of debris. Some deposits from violent events are now known to have a "low aspect ratio," which is a ratio of debris thickness to area covered. Finding thin deposits spread a great distance meant that while a relatively modest amount of material may have been ejected during an eruption, the force driving the eruption was a considerable hazard. Hoblitt said of the destroyed village's thin deposit, "One of the lessons that started to come home to the community after St. Helens is that these very energetic explosions produce deposits that are thin in comparison to their extent. The reason why that is significant is that if you're trying to read the history of a volcano by examining its ancient deposits, these thin deposits, which might easily be ignored or overlooked in preference to the thicker deposits, can have a great deal of significance as to the hazard potential of the volcano."

In fact, as Hoblitt and others were finding, events powerful enough to scorch and sandblast a ridge five miles away might in time leave no clue at all. Traces may disappear faster in Indonesia than they would in the Pacific Northwest because of the intensity of weathering and the large amount of crawling things in the soil. But it was becoming obvious to Hoblitt that while volcanic debris had to leave an honest record of an eruption, that record might not tell the whole story. Most important, because of soil churning, geologists could be easily fooled into believing they knew the reach of a volcano's punch. The lesson was that the discernible history of a volcano might not include the full range of courses a volcano could take.

By the time Hoblitt left Indonesia, he had experienced a buffet of volcanoes and studied their behaviors. In a few months, Hoblitt had gained experience on eruptions, hummock fields, seismology, base surges, dome collapse, and much more. In one field season, he had seen more volcanoes and more types of eruptions than perhaps any other Survey volcano specialist. He had learned not to trust so completely the debris record of an eruption. And perhaps most important, he had seen firsthand that low-energy eruptions can be more dangerous than extremely energetic explosions.

CHAPTER 9

After Armero

In its own way, Colombia's Nevado del Ruiz volcano, which stood at the northern tip of the Andes, would have as profound an impact on the Survey's volcano programs as Mount St. Helens had, although its eruption was a squeak compared to the May 18 blast. The events began in November 1984, when mountain climbers reported that the volcano's steam vents were unusually active. Earthquakes were felt at the Refugio lodge near the summit of the snowcapped mountain, and in late December, the snow was covered with a layer of fine ash and a yellowish sulfur veneer.[1] After more than a century, Ruiz was showing signs of fire in its belly.

The danger posed by Ruiz came not from pyroclastic flows or directed blasts, but from mudflows. Although Ruiz was just five degrees north of the equator, it was a 17,680-foot-high mountain. At that altitude, it was cloaked in a majestic cape of snow and ice, which is to say, stored water. In the past, mudflows had swept down the mountain without warning and buried families, farms, and entire towns. The last great tragedy the mountain unleashed was in 1845. Colombian naturalist Joaquin Acosta described the lahars that raced down the Lagunillas River in the journal of the Academy of Sciences of Paris. He wrote:

"Descending along the Lagunillas from its sources in the Nevado del Ruiz, came an immense flood of thick mud which rapidly filled the bed of the river and covered or swept away the trees and houses, burying men and animals. The entire population perished in the upper part and narrower parts of the Lagunillas valley. In the lower part, several people were saved by fleeing sideways to the heights; less happily, others were stranded on the summits of small hills from which it was impossible to save them before they died."

A hundred years after Acosta wrote that description, the devastated area was thriving. Mudflows rolling down from Ruiz century after century had left behind fantastically fertile ground. As memories of the 1845 mudflow faded and were replaced by more lucrative dreams, people began planting rice and coffee fields in the new earth. Then they built their homes on it, and eventually entire towns grew up on the old mudflows.

This is a standard story around many volcanoes. People everywhere are moving closer and closer to dangerous volcanoes, sometimes living right on the slopes of active mountains. They are drawn to them because the land is so productive and available to them that poor farmers put economic opportunity ahead of personal safety. Or they move there because the view is spectacular, as in the area south of Seattle, where entire communities now stand on mudflows as Armero did in the mid-1980s.

The town of Armero was a prosperous farming community that straddled a bend the Lagunillas River had carved out as it dropped to the Pacific. Armero's population of twenty-nine thousand lived on the 1845 mudflow, which was about thirty miles from Ruiz.[2]

For months after the first signs of awakening in November 1984, the volcano churned out sulfur plumes, but little was done to map the hazards, monitor the activity, or prepare for a disaster. In late March, a seismologist from the United Nations Disaster Relief Organization called the UN's Natural Hazards Unit in Paris and complained, "Nothing is being done about this new activity. It appears that the university in Bogotá does not have volcanic expertise, and neither does [Colombia's] institute for geology and mines."

By summer, the situation had changed little. In May, a scientist from Ecuador investigated on behalf of the UN and the World Organization of Volcano Observatories. He reported, "The crater's activity remains stable but in an abnormal state." He, too, complained that "no monitoring activities were being carried out. Portable seismographs, if they exist, are still in Bogotá." By the end of May, Colombia's Geology and Mines Bureau (INGEOMINAS) asked the U.S. Geological Survey for technical expertise and equipment. But with eruptions in Hawaii and St. Helens continuing, the Survey said they could not help. In June, the Survey shipped some geophones and cables, but it was still unable to send someone familiar with the monitoring equipment. In a note written to the UN's Disaster Relief Organization, the Survey's deputy chief for Latin American affairs wrote, "The opportunity is clear, and it is unfortunate that we can spare no one from the Hawaii or Cascades Observatories."

In response to UN pleas, the Swiss government did send one volcano specialist and three seismographs. In late June, the Colombian government was notified by letter that the UN had put together a volcano response package—including volcanologists, equipment, and a training program. All that was required was an official request from the Colombian government. Unfortunately, the letter with the government's response was lost for months. Still, by late August, a primitive seismic network was being deployed by two independent groups, although data had to be carried physically to Bogotá to be analyzed, a process that took days at best.

On September 11, at one-thirty in the afternoon, Ruiz released a strong steam eruption, which lasted for seven hours. By midafternoon, a small lahar had moved almost twelve miles down the west flank of the volcano. This did not impact Armero, which lay on its eastern slope. But the event, which caused only minor damage, sent the residents of Manizales into a state of "fear, if not panic," according to local news reports. With Manizales's population of 350,000 now feeling threatened, this city became a catalyst for action. Manizales was also the primary focus of attention, and lying to the west it diverted official attention away from smaller towns such as Armero on the volcano's east flank where historically lahars had been at their worst.

The September 11 eruption did trigger some action at USGS. An administrator from the headquarters office in Reston, Virginia, was dispatched to Colombia. Darrell Herd had actually studied Ruiz for his Ph.D. For a week beginning September 20, Herd worked in Colombia with geologists and the civil defense authorities to consider past eruptions as a guide to what Ruiz might hold in the future. This was pure Crandell/Mullineaux-style work. What the digging confirmed was that

Manizales was on a river at little risk of damage from a mudflow launched from the peak of Ruiz. On September 23, Herd gave a public lecture in Manizales to try to quell the anxiety in the city. He said mudflows were unlikely to affect populated areas on the volcano's west side like Manizales. But his remarks were interpreted by many as a statement that applied more broadly.

By the time Herd left on September 27, few monitoring or hazard-mapping problems had been solved. By late September, there was still no radio-linked seismograph on or around Ruiz. Data had to be retrieved by hand daily. The seismic information collected by INGEOM-INAS then had to be mailed to the National University of Bogotá for interpretation. Even then, the data was not freely shared with the Risk Committee that had been established in Manizales to help monitor the volcano and prepare for a possible eruption. In fact, no seismic information was issued by INGEOMINAS until October 7.

A preliminary hazard map, which plotted regions that were vulnerable to a lahar and other features of an eruption, was finally issued October 7. As hazard maps commonly do, this one raised many concerns. Even before it was unveiled, the Manizales Chamber of Commerce warned that irresponsible reporting of hazard areas would "cause economic losses." The local archbishop said it was "volcanic terrorism." The magazine *La Patria* said in its cover story on the volcano that the map would cause "real estate devaluation." Government officials in Bogotá criticized it as "being too alarming."

The preliminary hazard report issued along with the map on October 7 by INGEOMINAS warned that the probability for a lahar along the rivers on the volcano's east side was 100 percent. The size of the mudflow would depend on the size of the eruption. But, it promised, "Armero could be evacuated in two hours."

The report created confusion. Residents in Armero didn't know if there was no danger to them, as had been reported to the citizens of Manizales, or if there was a virtual certainty that mudflows would come with any eruption. And while some felt comforted by the assurance that there would be a two-hour warning, allowing people time to move to safety, others were uncertain if the responsible officials would err on the side of caution rather than trigger a mass evacuation over a false alarm.

Meanwhile Ruiz continued to act menacingly.

After watching the disorder from afar, the U.S. Geological Survey finally proposed sending six telemetered seismographs to Ruiz along with a seismologist who was becoming known as the "Indiana Jones of volcanology," Dave Harlow. Harlow was a former Vietnam Marine. After joining the Survey, he had worked primarily on threatening volcanoes in Central America. By the mid-1980s, he had become the Survey's most experienced seismologist working explosive volcanoes.

The Survey sought financing for the project from the State Department's Office of Foreign Disaster Assistance. Weeks passed as the State Department tried to negotiate a lower budget for the project, but by month's end, Harlow was preparing to fly to Colombia on November 7.

It never happened. Before Harlow's departure, Colombia was shaken when guerrillas attacked and occupied the Palace of Justice in Bogotá. Colombia's president refused to negotiate with the leftist rebels and ordered government troops to retake the courthouse. After repeated assaults with armored cars and troops using heavy weapons and explosives, the Palace of Justice was liberated. But in the battle, one hundred people died, including eleven supreme court justices.[3] In the wake of the violence, the State Department decided not to send nonessential U.S. government employees to a volcano just 105 miles northwest of Bogotá.[4]

Although INGEOMINAS had certified that Armero could be evacuated in two hours without danger, the hazard map, which was supposedly the foundation for the judgment, showed no safe refuge *in* Armero. In fact, the nearest spot that could be identified as beyond the hazard zone was about a mile away. Evacuating a town of twenty-nine thousand people to a refuge over a mile away would be difficult. If it was a false alarm, it would incur enormous political as well as financial costs. But the hazard map was a compelling argument to err on the side of caution, since it identified areas around Armero that had been buried under as much as twenty-six feet of mud by the 1845 lahar.

But the main problems were, first, if the town could be evacuated in two hours, who was going to warn the people to start, and second, did anybody really know what to do if such a message came?

On November 13, just one day after the revised hazard map was supposed to be unveiled, Armero ran out of time. Shortly after 3 P.M., as a meeting of Red Cross and civil defense planners began in a town far south of the volcano, Ruiz began shaking and spouting ash. The civil defense group was notified of "mild" but continuous activity at Ruiz, although no one knew what "mild" actually meant. They ordered a message sent to the river cities, including Armero, on the east side to "sound the alarm—*if necessary.*" For six hours, the mountain continued to sputter. No threatening mudflows were sighted and no alarm was sounded. In fact, the people listening to Radio Armero heard only music and repeated reassuring messages, which were echoed by the local priest.

At 9:08 P.M., Ruiz changed. The eruption turned magmatic, and the volcano began spewing hot ash. Later analysis would show that this lethal eruption was a minor outburst by volcano standards. It produced just one-tenth the material of St. Helens's May 18 blast.

The pyroclastic material coughed out had temperatures that reached 1,650 degrees Fahrenheit. The searing ash and pumice quickly melted a broad swath of snowpack and glacier of Ruiz's eastern face. The high-mountain flood of hot ash, rock debris, and melted water began channeling into chutes. Soon the hot waters were cascading down the steep slopes at speeds approaching one hundred miles an hour. The flood sheared off slabs of ice and scoured the earth down to bedrock in some places. The force of the flow was such that it picked up sections of forests and boulders as big as buses. To those nearby, this thing eating the landscape made a deafening roar. Seven minutes after the start of the eruption, mudflows were already avalanching into the headwaters of the Azufrado River. As they raced downstream, they scraped out canyon walls and river bottoms until the mudflow, which looked as fluid as river rapids, but yellow, was the consistency of wet concrete. The mudflows of the Azufrado emptied into the Lagunillas, and the churning mass began a course that dropped three vertical miles to Armero.

Later calculations revealed the wave of mud rushing down the Langunillas was equivalent to the instantaneous release of the reservoir behind the world's largest arch dam. Occasionally, the flow would be held back as debris piled up at a bridge. But all the bridges eventually gave way after being pounded by boulders and trees bigger than church steeples. These brief delays made the mudflow begin to move downstream in waves. At 10:30 P.M., a relatively small mudflow rolled over the lowlands near a village named Chinchina, destroying three bridges and two hundred houses, and suffocating or crushing eleven hundred people in a little under ten minutes.

The main lahar was still an hour away from Armero, where people were preparing for bed, many of them listening to the music from the city's radio station. Colombian civil defense authorities in another city, who had been informed that a major lahar was on the move, at last decided to order the evacuation of Armero. But power outages and communications problems prevented the message from getting through.

A few miles upstream from the city, the mudflow muscled its way through a natural dam, which collapsed and added even more water and debris to the flow. When concerns were raised about an eruption, Armero's mayor was on the radio saying "not to worry, that it was a rain of ash, to stay calm in our houses."

At 11:35 P.M., the leading edge of the lahar, now 130 feet high, burst through the canyon just above Armero with a roar like a swarm of fighter jets. Above the canyon, the thundering flow shook the ground so violently that farmer José Rojas was thrown from his bed. He grabbed a flashlight and ran outside and saw a wall of water coming down the canyon and flooding out into Armero.[5]

As it broke out of the canyon, which was a natural funnel for the flow, the mud fanned out. The lahar's height dropped to thirty feet and its speed slowed to twenty-five miles per hour—still higher than any building in Armero, and faster than anyone could run.

Armero resident Hortensia Oliveros, nineteen years old and in her eighth month of pregnancy, was awakened when she heard screams in the street. She woke her mother, husband, and eleven-month-old daughter and they raced into the street. "We got about a block and a big wave of mud hit us," she said later. "We all went tumbling. The baby was knocked out of my husband's arms, and we were all separated." The young woman was swept under the mud and churned like laundry. When she surfaced, she saw the lights of the town had been knocked out. Her husband was yelling to her to try to grab on to a tree to stay afloat. She did, and a moment later the tree swept her away and her family disappeared forever.[6]

As the mass swept through town, it was so powerful that it cut down building after building and leveled the city's tallest and perhaps strongest structure, the church. In the darkness of those first few minutes, perhaps ten thousand people were swept into the lahar.

As the mudflow crashed through the center of town, a group of geology students staying at the hotel ran into the street and saw a wall of mud illuminated by cars that were churning in the flow. They raced back into their hotel and braced themselves. The solid concrete structure with thick walls was pounded again and again, and soon the rear of the building gave way. Then the entire structure collapsed. The students were thrown into the flow, but unlike in a normal flood, boulders and chunks of concrete were suspended in the rolling mass.

"Since the building was made of cement," said one of the students, "I thought that it would resist, but [the boulder-filled mud] was coming in such an overwhelming way, like a wall of tractors, razing the city, razing everything. [Then] the university bus that was in a parking lot next to the hotel was higher than us, on fire, and it exploded, so I covered my face thinking, 'This is where I die a horrible death.' There was a little girl who I thought was decapitated, but what happened was that her head was buried in the mud. A lady told me, 'Look, that girl moved her leg.' Then I moved toward her and my legs sank into the mud, which was hot but not burning, and I started to get the little girl out, but when I saw her hair was caught, that seemed to me the most unfair thing in the world."

The thick, rocky flood rolled over Armero in waves for almost two hours. The rushing mud never subsided to less than six feet, and in some places never under fifteen feet. The final, still tide of mud was soft and soupy. People not killed by the churning rock and tree mixture were left suspended in mud, a few up to their necks, unable to move.

"It started to be light, and that's when we lost control," said one of geology students, "because we saw that horrible sea of mud, which was so gigantic. There were people buried, calling out, calling for help, and if one tried to go to them, one would sink into the mud. So now you must start counting time as before Armero and after Armero."

In the morning, a few rescue helicopters arrived and the crews saw that what had been Armero was buried under ten to fifteen feet of mud. Of the town's forty-two hundred buildings, only the walls of a few dozen were still standing.[7]

In the dawn, helicopters began lifting a hundred survivors from hilltops where they had been stranded. At least another hundred people were trapped alive in the fetid mud. Rescuers were unable to reach most of them since the mud, although no longer moving, was like a vast field of quicksand. Three days later, dozens of people were still alive in the mud. One of them, a twelve-year-old girl, was wedged under her aunt's body. Only her head was above the mud. She died on Saturday, calling for her mother. On that same day, Dr. Ferdnando Posada, who had worked round the clock for three days trying to free those who were trapped, said, "The best thing that could happen here is another avalanche to cover those who are still alive."[8]

In Armero alone, twenty-one thousand people died and another five thousand were injured. The dead were left in place. Armero was no longer a city but a cemetery.

Armero would provide many lessons for volcanologists. Analysts were united in the assessment that Ruiz should have killed no one. Colombian officials had plenty of warning from the volcano that a potential disaster was brewing. Although the hazard maps were not developed until late and even then poorly circulated, geologists had identified those places at greatest risk. They correctly identified Armero as vulnerable from a mudflow. They even had short-term warnings on November 13.

Barry Voight, the landslide expert from Penn State who had studied the landslide potential at St. Helens, wrote what became a classic paper analyzing the failures at Ruiz. His conclusions crystallized the problem at the heart of the Armero disaster and, at the same time, outlined the problem that volcanologists would now confront at every volcano crisis. It was a new problem that had developed precisely because of volcanologists' growing ability to better—but not perfectly—call a warning. And that problem became the biggest remaining hurdle confronting the effective use of the science.

"At Ruiz," Voight wrote in his 1990 paper in the *Journal of Volcanology and Geothermal Research*,[9] "it was not so much the imprecision of science, or inept last-minute decision-making, or breakdowns of a few communication systems at the vital moment—the failure was to *wait* until the last possible minute. One cannot expect emergency management to operate efficiently at that time scale; but this is often what human nature seems to demand. It seems less a matter of excessive confidence or false sense of security than *the lack of will to act in the face of uncertainty* [my emphasis] and the unwillingness to accept the costs implied by the finite probability of a false alarm."

This lack of will, Voight was saying, was the cause of death of twenty-one thousand people in Armero alone. If he had stopped there, his paper would have been a significant contribution to the field. But he went further to calculate the real costs associated with issuing a false alarm.

He noted that the cost of calling for an evacuation would have been high. Security against looting would have been difficult, and leftist insurgents might have tried to exploit the crisis. And if it proved to be a false alarm, the population of Armero would likely have resented being ordered from their homes and having their livelihoods disrupted. Wrote Voight, "The authorities on the whole acted not unreasonably but were unwilling to bear the economic or political costs of early evacuation or a false alarm."

Voight's assessment was an earthquake in volcanology. Many geologists interpreted his paper as an indictment of the scientists for their failure to communicate the appropriate level of risk Ruiz posed.

Geologists advising civil authorities anywhere about a local volcano were now between a rock and a hard place, just as they had been in Long Valley. The scientists were learning how to better identify when a volcano was nearing an eruption. But their predictions about when it might actually pop were never going to be 100 percent accurate because of the complexities of nature. Their estimates would always contain a margin of error. This added a new dimension to the job of an expert adviser. It meant the scientist had to be willing to accept the consequences of vigorously advocating a massive evacuation, only to have nothing happen. *They also had to accept the consequences of not encouraging a needed evacuation forcefully enough.*

Voight was saying that if geologists want to be discreet advisers, that's fine. But when they walked into a volcano crisis, their powers to better monitor and assess hazards now came with a greater responsibility to vigorously advocate a course of action based on what the mountain was telling them. The wiggle room—that they were only advising local decision makers—was rapidly vanishing.

Voight also found plenty to criticize specifically about the U.S. response. Primarily, he observed that despite requests for help, experts knowledgeable in monitoring, hazard assessment, and even the local conditions at Ruiz never arrived in a timely fashion. In fact, the Survey sent only one person to the site before the eruption, and for only a few days.

"In part," wrote Voight, "this may reflect bureaucratic sluggishness or the reluctance of administrators in such agencies as Survey and

OFDA [the State Department's Office of Foreign Disaster Assistance] to release personnel or provide *predisaster* support for foreign assignments. In any case, this factor contributed to the delays in production of the Ruiz hazard maps and in effective monitoring."

Even before Voight published his autopsy of the Ruiz disaster the State Department had identified a failure in its own guidelines: that disaster assistance generally meant postdisaster. With this recognition, the State Department began an exploration of the possibilities of intervening *before* a natural disaster occurs overseas. Perhaps it was possible to actually reduce the threat or mitigate the destruction by predisaster intervention. At the time, the State Department was under increased pressure to do more with less. So, the idea took hold in the Office of Foreign Disaster Assistance that spending a small amount up front to mitigate a disaster could reduce the huge cost in humanitarian aid following a disaster.

Volcanoes became a natural focus of the new OFDA effort. The majority of the world's most dangerous volcanoes are in developing countries. And because of population growth and the attendant pressure to utilize all available land, the dangers from these volcanoes were becoming greater with every passing year. For example, the surge and mudflows from the Mayon volcano in the Philippines killed twelve hundred people in the 1800s, but the population at risk in the same danger zone was approaching more than three hundred thousand at the end of the 1980s.[10]

With this in mind, the State Department called the Survey and said they were ready to work with Norm Banks.

In December 1980, seven months after St. Helens had erupted, a memo was circulated among the observatory staffs in Hawaii and Washington State that suggested it was time to pull back, think about where the Survey's volcano programs should go, and to suggest "all possible (and reasonable) areas of research." In January 1981, one of the responders, Norman Banks, suggested building a team of volcanologists who would operate like "smoke jumpers," responding to every volcano crisis around the world (assuming they were invited). During the early days of the St. Helens crisis, Banks had paid his own way from HVO just to work at the mountain and wound up assisting Jim Moore in shooting deformation lines. Several other people had the same idea at about the same time, including Rick Hoblitt and Dave Harlow, but

Norm Banks was the first to put it in writing and became its tireless advocate.

In 1980, Banks recognized that if the Survey was confined to the United States, it might be another fifty years before U.S. volcanologists got the kinds of valuable experiences they had acquired at St. Helens. If they waited that long, the human expertise the Survey now had would have died out and the technology would never be pushed.

The Banks proposal had two goals. The first objective was "to provide to requesting nations an on-the-ready, experienced, and well-equipped team for observations and hazard evaluations of erupting and stirring volcanoes." But it would also serve the Survey by increasing "the experience, the equipment inventory, and the quality and innovation base for observation, interpretation, and prediction of volcanoes."

He envisioned a standing team of ten experts and a stash of state-of-the-art monitoring devices. Essentially, the experts and their equipment would be a portable volcano observatory. This would be a self-contained team, ready to go anywhere when the bell rang. He proposed the team have a leader, who, among other duties, would maintain a library on the world's most dangerous volcanoes; a machinist-electronics expert to develop and maintain equipment; two photographers; and a flock of "current or past members of observatory staffs" for on-site monitoring, evaluating, and advising.

Banks also envisioned another feature to his SWAT team: he planned to leave the entire cache of monitoring equipment on the volcano under study and train local geologists to evaluate the hazards, run the observatory, and interpret the data once his group pulled out. In the big scheme of things, Banks was planning to create a network of well-monitored volcanoes around the world. Data from this network could, he felt, move the field of volcano science at warp speed.

The start-up cost for such a crisis response team, he estimated, approached $200,000. His proposal was never turned down officially, but it was put in the pile of projects considered for support if everything else was funded. Banks, who, like many scientists, was always intense about his ideas yet too often blind about the impact he had on people, nevertheless pressed his case at every opportunity. As he lobbied, he began stockpiling monitoring equipment.

In May 1981, the Pagan volcano just north of Saipan in the Mariana Islands in the western Pacific erupted. Banks pressed Survey managers, and five days after the eruption began, he and two colleagues had deployed equipment to read the volcano's quakes, tilt, and gas emissions. The experience was a challenge. Equipment was lost en route. Local bureaucracies in the Marianas delayed the team's flight for nearly four days. He struggled with a minimal budget for supplies, air freight, and local transportation. Most of all, the team was too small and worked itself into exhaustion within eight days. But by the time they had to leave, the team had harvested enough data to publish a paper on Pagan. The paper helped win supporters among Survey officials, who began to agree that a crisis team could be productive. But it didn't change the reality that funds didn't exist to give it life. So, Banks pushed his concept whenever he heard a volcano showing signs of renewed activity.

Banks would continue to refine his plan and submitted his funding proposal for a Volcano Crisis Assistance Team (VCAT) in 1982, 1983, and 1984. Finally the State Department's US AID office agreed to support the program if the Survey's Office of International Geology could match the funds. The Survey couldn't. Banks then began looking for funds elsewhere and applied to the National Geographic for a $100,000 grant. The National Geographic was tentatively interested, but again, the Survey torpedoed that idea.

Then in late 1983, a huge volcano system in Papua New Guinea (PNG) began showing signs of unrest. Rabaul was a monster caldera, about as big as Long Valley in California, and it had frequent eruptions, the last one during World War II. Tens of thousands of people now lived on the ring of the gigantic caldera, so the potential existed for a catastrophe.

In January 1984, the State Department's US AID Office of Foreign Disaster Assistance began receiving cables from the PNG government requesting technical assistance for Rabaul. State called Survey officials, who eventually tapped Norm Banks for an estimate of costs to provide that assistance. But concern for Rabaul faded as the volcano's activity decreased, and Banks's proposal faded with it.

After Armero, however, the Office of Foreign Disaster Assistance, which would spend $3 million to help the victims of Ruiz, began thinking a $200,000 insurance policy was a good idea after all. A five-year experiment was approved, the costs to be split between OFDA and the

Survey. The team became known as the Volcano Disaster Assistance Program or VDAP.

The program was launched at an interesting time. Volcano monitoring equipment was being miniaturized and modernized. The computer revolution was unfolding and gave field volcanologists new and more powerful monitoring tools. A new software package had just been designed by a famous Survey programmer, Willie Lee. The "Willie Lee system," as it became known, allowed for real-time acquisition and analysis of seismic data on portable computers in the field.

VDAP was sent to volcanoes in Guatemala, Mexico, Bolivia, Ecuador, Argentina, Peru, and Chile.

Soon, however, some regular team members such as seismologist David Harlow began to draw pointed criticism from superiors for their participation in such international fieldwork. In annual evaluations, many team members began to receive slow promotions or be denied pay increases because they were spending more time chasing volcanoes than writing scientific papers. Team members replied that they were providing a true benefit to people living in the shadow of volcanoes, but that defense did not carry much weight.

Eventually, a conflict arose within the Survey between the "meatballs" and the "coneheads." The meatballs, personified by Dave Harlow, were irresistibly drawn to volcano crises. These were the people who loved flying off to the jungle, setting up a monitoring network on rocking volcanoes, and guiding locals into mitigation efforts. The coneheads were their more academically minded brethren who worked on formation temperatures of magma and the subtle geochemistry of lava, basic stuff. It is a common cultural divide in science—basic versus applied science. The basic researches—such as Jim Moore and Don Swanson at St. Helens—were driven by the pure joy of uncovering the fundamentals of nature. The applied scientists—including Rocky Crandell and his descendants—believed that making the discoveries of basic science useful to society ennobled their work.

In fact, one really isn't of much use without the other. But, just as a clash of cultures had produced bitterness at St. Helens, this reincarnated form also turned the atmosphere sour in the Survey's volcano program. The difference this time was that virtually every supervisor was a conehead. So almost every question involving the pay and promotion of one of the meatballs was resolved by an evaluation team of coneheads, often unfavorably for the meatball.

At the same time, Norm Banks was proving to be a better inventor than a manager. In his years of lobbying for the VDAP concept, Banks had left a trail of strained and broken relationships. His style of leadership was just as awkward, and scientists working under him grew increasingly unhappy. Though it was painful, Banks resigned as the VDAP chief in January 1991. Dick Janda replaced him and revived the program. Eventually, the job of VDAP chief went to Dan Miller.

After Armero, geologists also had an audience prepared to listen to their warnings of the possibility of potentially gigantic lahars in the United States. About an hour's drive north of the Cascades Volcano Observatory was perched America's most dangerous potential lahar, on the snowcapped slopes of the mountain that looms serenely over Seattle and Tacoma, Washington State's Mount Rainier.

Rainier is one of the most interesting volcanoes in the United States. Its snowy shoulders top out at 14,410 feet above Puget Sound, and it has a long history of eruptions, the last one occurring just 150 years ago. It is at Rainier's summit that climbers realize they are standing on a volcano. Steam vents ring the summit crater. And the ground is so warm that, even at this altitude, a few places on the summit are bare or covered only with a thin layer of new snow. Snow caves formed by the rising steam near the summit are a refuge for stranded climbers. Surface temperatures taken at the summit have registered as high as 176 degrees Fahrenheit.[11] The heat is from the magma in Rainier's belly.

Rick Hoblitt's boss in the Survey, Rocky Crandell, was the first to recognize that Rainier had launched massive lahars. Back in 1953, in one of his early assignments with the Survey, Crandell was assigned to a team preparing geological maps for the Puget Sound lowlands, at the foot of Rainier. He immediately became fascinated by their geology. He saw from highway cuts that the area had been repeatedly blanketed by glaciers.

During Crandell's first summer of mapping, he encountered a geological puzzle as big as a billboard. It was displayed on a hillside that had been cut over several thousand years by the White River. Land that has been sliced open by rivers or road builders exposes the earth in ways that are usually hidden from geologists walking on the land's surface. These cross sections reveal geological history. They display layers of the earth's evolution, like vertical tree rings, and each layer had its own story to tell. Crandell became an expert in deciphering these stories.

The cut Crandell saw was perplexing. The top of the cut was flat, like much of the surrounding terrain. But looking deeper at this cutaway, he could see the buried, earlier features that were made up of small hills. But something had come along and filled in the spaces between the hillocks. The fill was just a few feet thick at the top of the buried hills and fifty feet deep in the hidden valleys. It had left a deceptively flat top, like the surface of a pond that hides boulders below.

Earlier maps had identified the region as a moraine, which is a land-fill dumped by a glacier. But that didn't make any sense to Crandell. A glacier would likely have been a few thousand feet to a mile thick. Such a weight would have eroded those small hills. Moreover, the material that filled in the hills was bouldery clay from pebbles to blocks as big as a house. Then Crandell learned that the Survey geologist who was mapping the block next to his own had discovered similar clay deposits. The two began working together and soon found that the generally flat low-land region was filled with these clay deposits that had been plastered over the countryside.

One night, in the summer home near Seattle he had rented for him-self, his wife, Mackie, and their three children, Rocky Crandell started a list of possible explanations. This list-making is a standard tool for geol-ogists. After the list is made, each explanation is examined to see if it fits the available data. One after another, the possibilities are crossed off leaving "the last man standing" as the most likely. But when Crandell was done, all the reasonable possibilities had been eliminated. Puzzled, Crandell wondered if the clay had *flowed* into place. Right then he drew a lightbulb in his notebook. Could it have been a monstrous avalanche of flowing clay? In time, Crandell and the neighboring geologist discov-ered the mudflow had once been the entire summit of Mount Rainier, which had collapsed. The mudflow had released three hundred square kilometers of material that had traveled seventy miles to Puget Sound. When it stopped, the mudflow had buried the original landscape, leav-ing a new, flat surface.

The Rainier mudflow Rocky Crandell found may not have been trig-gered by an eruption, and any future mudflow from Rainier might not need an eruption either. The mechanism was different from that seen at Armero and even at St. Helens.

Volcanoes can sometimes make what some geologists casually call "rotten rock." Over hundreds of years, volcanoes heat the acidic water that drains through the mountain, and this water transforms the rock to clay. Clay is particularly unstable. Eventually, large portions of volcanoes have been cooked into clay. Then, any minor eruption or perhaps even a minor earthquake is all that is needed to release a mudflow.

About fifty-seven hundred years ago, something, perhaps an eruption or maybe just an earthquake, loosened a mile-wide slab of Rainier's acid-softened rock, removing two thousand feet from its summit. Sliding down the mountain's east slope, the warm clay mixed with snow and stream water and was transformed into a mass of mud sixty times more voluminous than the lahar that buried Armero. It surged down three valleys at speeds over one hundred miles an hour and flowed into the flatlands spreading over an area of 120 square kilometers.[12] There, its speed slowed to under thirty miles an hour but it still had the force to carry away boulders and forests with trees two hundred feet tall. The mass flowed over land where six towns now exist and finally dumped into Puget Sound near what is now Tacoma.[13] This was known as the Osceola Mudflow.

Crandell's lahar work was a major contribution to geology in general and to volcanology in particular.

Crandell and others went on to find more than sixty other mudflows from Rainier in the last ten thousand years. A few of them ran out seventy miles to Puget Sound. Along the way, he also realized that if the Osceola Mudflow had been unleashed in the mid-1950s, it would have covered entire towns without warning. By the late 1980s, geologists were becoming concerned that the United States might have its own Armero to worry about.

This concern triggered a study published by the prestigious National Research Council that concluded in part: "Major edifice failures, glacier outburst floods, and lahars could occur in the absence of volcanic eruptions because of the inherent instability of the edifice. Mount Rainier is a high volcano (14,000 feet above sea level and approximately 9,800 feet in relief) that contains about 140 cubic kilometers of structurally weak and locally altered rock capped by about 4.4 cubic kilometers of snow and ice, all of which stand near the angle of repose. Ground shaking during an earthquake, or ground deformation due to intrusion of magma into the edifice, could cause the gravitation failure of a large sector of the volcano, producing catastrophic avalanches and debris flows and possibly triggering an eruption. Glacier outburst floods and lahars can also occur during heavy rainfalls or transient heating events that melt snow and ice on the volcano. . . . Damage caused by debris flows could be substantial."[14]

Because the volcano is so high and the Pacific Northwest is so wet, Rainier is perpetually draped in a vast amount of ice and snow. The mountain has twenty-six named glaciers. All told, the volcano carries thirty-six square miles of the ice and snow, more than twenty times that of pre-eruptive St. Helens. When it comes to dangerous volcanoes in the United States, Rainier's combination of snow, ice, and clay-rich interior puts it in a class by itself.

Another Ruiz-like lahar could be unleashed by a moderate earthquake or an eruption of pyroclastic debris over the snowpack. If it happened at night or on one of the many days when the volcano is wrapped in clouds, early warning time could be nil.

About one hundred thousand people live in homes built on Rainier's lahars. The town of Orting, for example, is built on twenty feet of debris deposited by the Electron Mudflow, which roared through the alpine valley five hundred years ago. Every day, commuters pass over mudflows that would have buried thousands of them without a trace.[15]

As the 1980s drew to a close, enormous strides had been made in American volcanology. There was no doubt that, with seismographs, deformation tools, and gas monitors, geologists had learned to tell when magma was moving underground. They had also learned to tell if magma was moving toward the surface.

The Survey had gained more experience on explosive volcanoes in Indonesia. The agency now had a fast-reaction crisis team that could deploy a volcano observatory anywhere in the world, often within days of the start of any activity. New technologies such as the Willie Lee system, which gave scientists real-time seismic information, were building the power and the rapid-analysis capability that volcano monitors could deploy. And Survey geologists were identifying the dangers from even quiet volcanoes such as Rainier.

The eruptions at St. Helens after May 18, 1980, had shown the Survey geologists that they could actually predict the time of an eruption, even if it was only a dome extrusion. The techniques used at St. Helens couldn't necessarily be transplanted as a whole to other volcanoes, as Mammoth Lakes had proved. But their success did make believers within the Survey itself. They recognized that volcanologists need not be totally baffled when confronting a threatening volcano. In fact, Survey scientists were growing confident that they could identify the precursors of an eruption. This technical expertise could help them guide officials responding to an emergency and save lives.

Still, if all these advances were to actually be useful, one more obstacle had to be overcome. It was the central failure at Armero: the lack of will of those responsible to act in the face of scientific uncertainty. The geologists were convinced that the scientific uncertainty would always exist. No volcano, not even St. Helens, had been so thoroughly understood that predictions were always right in all details. And most of the really dangerous volcanoes, especially those outside the United States, were so little understood that hazard assessments, for example, would have to be quick and dirty jobs.

But once they had a history of a volcano, however condensed, and when their field computers started to pick up a volcano's pulse, the geologists had to have the confidence of the local officials. Those officials needed to believe that the dangers their communities were facing were powerful and real.

This was the problem Chris Newhall chewed on.

Newhall had been one of the new hires the Survey was allowed right after the May 1980 eruption of St. Helens. He was then a newly minted Ph.D., but he had previous experience with explosive volcanoes in Guatemala. He had also served in the Peace Corps in the Philippines and had married a Filipina while there.

After Newhall was hired by the new Cascades Volcano Observatory, he was trained by Rocky Crandell and Don Mullineaux in hazard mapping. He became something of an expert himself in reading ash deposits. But his real duty at CVO was as hazards coordinator, the job that Crandell and Mullineaux had had during the crisis. Most of Newhall's work at that time involved taking regular readings from the other geologists, integrating their assessments, and translating that current picture of the volcano's activity for the Forest Service, the Washington State Department of Emergency Services, logging companies, and anyone else who had an interest in the hazard zones.

Accidentally, Newhall also continued the strained relationships Mullineaux and Crandell had had with Don Swanson. Newhall didn't set out to confront Swanson, but to him, Swanson's repeated dashes into the crater were not justified by the data he was bringing back. Newhall was just one of the people arguing the point, and one of many who lost the argument. But the debate, which extended over months, soured the atmosphere at CVO.

After being primed by Mullineaux and Crandell to focus on public safety during an eruption, Newhall could not ignore the tragedy at Armero. To his way of thinking, decision makers would always delay their evacuation orders until it was too late because they simply didn't understand the risks involved. If everyone involved with Armero had understood the power of a mudflow, Newhall was convinced, alarms would have been sounded and people would have walked a few hundred yards to higher ground.

"The words *flujos de lodo* [Spanish for "mudflow"] just didn't mean a thing to the people of Armero," he says. "Even if you scrape away and set aside all the other complications that occurred in Colombia at that time, such as the bombing of the Supreme Court and the guerrilla movements and the fact that there were multiple political jurisdictions the volcano was split into—those were clearly contributing factors. But I think the number one factor is that they just didn't have a clue about what was coming at them. If they had, then none of the other complications would have mattered. They would have dealt with it. But they just didn't. Your average person on the street, and I don't care whether you are in Armero, Colombia, or at the foot of most any other volcano that's been quiet for one hundred years, they have no memory at all and no concept of the kinds of things that can come at them."

This was a view shared by many volcano experts when they gathered in January 1987 for a celebration of HVO's seventy-fifth anniversary. While the event was a celebration of the work done at HVO, Armero hung over the festivities like the death of a relative.

A new building was dedicated at the HVO facility, and a lecture was given by Maurice Krafft. Maurice and his wife, Katia, were French filmmakers, photographers, and volcanologists. They traveled the world filming volcanoes erupting and returned to France to give lectures that paid for their next adventure. They often ignored official restrictions and simply charged off to an eruption to get the best possible film. Once they even found themselves in a rubber raft on a lake that was rich in sulfuric acid from venting magma gases. The scenes they filmed were the most powerful images of volcanoes ever taken.

For his lecture, Maurice had strung together his best clips. As Maurice narrated, the scientists sat mesmerized by scene after scene of the terrible power of eruptions.

As he watched the film, Newhall thought that if only the doomed people of Armero had seen some of this film, they might have grasped what was coming down the canyon. When the lecture ended, Newhall and former HVO scientist-in-charge Robert Decker approached Maurice in a hallway outside the auditorium.

The discussion continued the next day on the rocks of Coconut Island, near Hilo. Now Norm Banks and Bill Rose had been added to the group, as had Steve Brantley, a public information officer at CVO who was on a leave studying documentary filmmaking in Chicago.

"We all decided," remembered Newhall, "that one of the things that probably would help a lot would be if we could have a hard-hitting video that showed what each of the major hazards is, how fast it goes, how far it goes, and what it does when it hits you. Nothing fancy. No fancy story development. Just plain, 'Here are the facts guys!' "

Maurice agreed to supply film from his own collection free of charge, and Brantley along with another member of Banks's VDAP team, John Ewert, would put it together. The film would be narrated in English, but a Spanish version could easily be dubbed as well. It took a while, but eventually a budget of $70,000 was cobbled together from a half dozen agencies, including the UN.

The clips were sobering records of volcanic catastrophes. They included a gruesome gallery of Armero's victims. The film also included pictures of forests rapidly churning in mudflows, bridges buckling, and houses carried away whole. It showed the problems associated with large ashfalls, including stark pictures of orchards turned gray and stripped of leaves, and rows of houses with drifts of ash up to their porches. The clips also included dramatic shots of eruption-driven tsunamis and the heavier-than-air bubbles of carbon dioxide gas that can asphyxiate animals and people.

By 1991, the film project was nearly finished. Many felt it was the last element to complete the mobile observatory. The monitoring equipment was now lightweight, completely portable, and provided a more detailed and more integrated view of a volcano's many vital signs than had been available in 1980 at St. Helens. More important, with VDAP's growing experience in Central and South America, the Survey was building a squad of scientists who were familiar with explosive volcanoes. In other words, the scientists now had improved ways of monitoring a volcano and a deeper background in understanding what their readings meant. With the Krafft film, they thought they at last had a potent tool to educate decision makers and local populations about the magnitude and types of dangers facing them. Soon, Pinatubo would test it all.

Field Notes

DECEMBER 15, 1989
REDOUBT, ALASKA

Shortly after 10 A.M., as the winter sun was just rising over Anchorage, Tom Miller started the morning shift at the headquarters of the Survey's new Alaska Volcano Observatory. After St. Helens, the Survey came to realize that they had a garden of explosive volcanoes in the United States. They were in Alaska, mostly growing in a line along the Aleutians. They made good specimens for studying the mechanics of explosive volcanoes. But the Survey also found it important to monitor eruptions because of a recently appreciated threat volcanoes pose to commercial jets—they gum up engines causing them to stall.

For two months, AVO scientists had been monitoring a growing intensification of quakes under Redoubt, a volcano 110 miles southwest of Anchorage. It took the Survey team a few weeks to agree that the quakes were new and real. The team had just established a seismic net on Redoubt, which had been coupled with the new Willie Lee system. The day the Survey team turned on the system, they saw a scattering of long-period quakes. One interpretation was that Redoubt always had these particular quakes, which had been associated with moving magma. Or perhaps Redoubt had just started to move toward a catastrophic eruption, as it had twice in the last one hundred years.

At 10:15 A.M., the seismic instruments began jumping. The swings were so wide that it was obvious the volcano was erupting. Redoubt had popped eruptions twice that morning, but this one was far more vigorous than anything else seen so far. Miller began working the calldown list. He first contacted the regional offices of the Federal Aviation Administration.

Since the first St. Helens eruption almost a decade before, volcanologists had learned to appreciate the special dangers that volcanoes pose to jets, even jets at a great distance from an erupting volcano. As pilots learned while flying through ash clouds, the grit can sandpaper windows and fry the avionics that help pilots fly their planes. But the greatest problem is that jet engines suck in a huge amount of air, and if that air is laced with volcanic ash, the hot engines can turn the ash to glass that will clog the turbine inlets and cause engines to shut down. Several jets had lost an engine from such encounters, and the FAA was especially concerned about the vulnerability of trans-Pacific flights, which routinely fly over some of the most volcanic regions on the planet. Each day, twenty thousand people passed over Anchorage or stopped there to have their planes refueled.

While the Survey was convinced that volcanoes posed real dangers to commercial aircraft, the FAA and commercial airlines were only dimly sensitive to the risks. In fact, in an analysis of the events of December 15, scientists were surprised to learn that in interviews with officials at Anchorage International Airport "none of those . . . could recall having faced the problem of volcanic hazards from ash clouds in the past."[1]

Shortly after 11 A.M., KLM flight 867, a new Boeing 747, began descending into Anchorage. The flight was from Amsterdam to Tokyo with 231 passengers on board and a crew of 14. Before departing Amsterdam, the crew had been briefed that Redoubt had been active, and in response the captain ordered an additional five thousand gallons of fuel in case the flight was forced to divert from Anchorage.

As the flight descended, a Boeing 727 passenger jet had just taken off from Anchorage and was climbing through eighteen thousand feet when the pilot radioed he had encountered an ash cloud and was being forced to return to the airport.

The 747's flying pilot (there were three pilots on board but the captain was not in command at that moment) turned southeast to avoid the cloud. No one on the flight deck could see an ash plume. The morning's first light illuminated only a few thin, white clouds to the south.

At 11:32 A.M., the flight crew was surprised when the plane slid from sunlight into darkness. Suddenly, the only thing the crew could see out the windows were objects that looked like fireflies. The cockpit became smoky and the plane's electronic instruments began failing. The pilot and her copilot put on oxygen masks as they began a full-power thrust to climb out of the cloud.

The 747 had flown into the ash plume at twenty-five thousand feet. Ninety seconds after starting the power climb, after a gain of twenty-nine hundred feet, all four engines stalled and the plane began to fall.

As the plane stalled, objects inside the passenger cabin began to float. Smoke started filling the cabin. Passengers began sensing a strong odor of sulfur.

On the flight deck, the pilot and her copilot were grappling with an entirely new in-flight emergency. No one had ever been trained to fly a 747 with all four engines out. According to the manufacturer, it could never happen.

The KLM 747 became a glider at 27,900 feet. The pilots had few options. Essentially, all they could do was check for the source of the malfunctions and try to restart the engines. But the engines would not restart, and the plane continued to fall at a rate of 1,620 feet per minute. It passed through 25,000 feet and then 20,000 feet as the pilots repeated the restart sequences. At 18,000 feet, still without power, the plane was just 7,000 feet from the highest peaks.

After falling for more than six and half minutes, two engines restarted at 17,200 feet, and at 13,300 feet the remaining two engines kicked over. The plane landed at Anchorage at 12:25 P.M., fifteen minutes late. Upon inspection, all four engines were found to be so badly damaged each had to be replaced. Extensive abrasion was found on the cockpit windshields, the wings' leading edges, the tail rudder, and the engine cowlings. Ash had plugged the pilot-Static system, which measures the aircraft speed while in flight, and ash had contaminated the engine oil, hydraulic fluid, and water supply. The bill to repair the $140-million plane was put at $80 million.

Redoubt kept erupting and its ash drifted south across Canada and the United States and into Mexico. On December 17, Redoubt ash clogged the engines of a descending 727 and shut one down. That incident happened near El Paso, Texas. Geologists and pilots alike were learning that eruptions had a long reach.

Part III

Mount Pinatubo, 1991

CHAPTER 10

Trained Decision Makers

In the spring of 1991, the man whom Rocky Crandell had hired to replace him at St. Helens was working at U.S. Geological Survey headquarters, in Reston, Virginia. Chris Newhall was then the deputy chief of the Igneous and Geothermal Processes Branch. It was his turn to labor as a manager. He worked in a rarefied section of the complex known as "carpet land" and had a pleasant view of the surrounding Reston woods. Newhall had about served his sentence; he had only six months left before he would be freed to return to real science. Lately, he had been preoccupied with organizing the Survey's first conference on the risk volcanoes posed to aircraft. He had no particular authority to convene such a meeting, but as he said, "sometimes when you call a meeting, other people come." It was set for July in Seattle, and it looked as if it would be crowded with aircraft and engine manufacturers, representatives from the commercial pilots union, the Weather Service, the Federal Aviation Administration, and ash experts.

Shortly after arriving at work on the morning of April 7, Newhall received a call from an old friend, Ray Punongbayan, who was then the head of the Philippine Institute of Volcanology and Seismology, or PHIVOLCS as it was called. Nothing specific was on Punongbayan's mind. He was just updating Newhall on recent activity, as he did a few times a year. PHIVOLCS was spread thin, Punongbayan told Newhall.

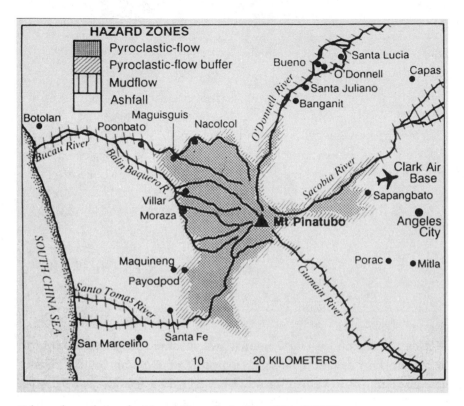

Volcano hazard map for Mount Pinatubo in May 1991. (USGS)

Many of his most experienced people were studying outside the country, and there was some unrest at Taal, a volcano just south of Manila. "Oh, by the way, Pinatubo is also showing signs of restlessness," Punong-bayan added. Pinatubo had popped some steam explosions on April 2 that had opened a fissure a half mile long. Along the fissure, three vents continued to steam. Punongbayan had responded by sending a team with seismometers to the volcano, and they had been picking up tremors from little earthquakes. There was not much more information than that.

Newhall had a long-standing interest in the Philippines, which began when he was assigned to the country by the Peace Corps in 1970. He had learned to speak Tagalog, married a Filipina schoolteacher, and worked with Punongbayan off and on for nearly twenty years.

Newhall's friendship with Punongbayan was cemented in 1984, when they worked on Mayon, a beautiful, near-perfect cone that rises straight from the sea. In early September 1984, Mayon had burst with a moderate eruption, but with little loss of life, and the volcano quickly began to settle down. Punongbayan, who was then chief of PHIVOLCS, and Newhall, who was working with him as a Survey adviser, were then pressured to allow the seventy thousand evacuees to return to their homes. The two geologists resisted. The eruption had coincided with the fortnightly tide. A theory had been circulating that when a volcano is teetering between eruption and quiescence, a minor gravitational pull, such as a fortnightly tide, can tip the balance. So, despite the growing clamor from the evacuees, Punongbayan, with Newhall's support, held them back. Newhall saw that extending the evacuation was a hardship for those displaced by the decision. It was a lesson to Newhall that determining when a volcano *won't* erupt can be just as important as predicting when it will. The evacuees lost income and many had property stolen. But the scientists wanted to wait until the next tide to see if it would trigger another eruption, and sure enough, it did. Mayon released a bigger eruption than the first. It lasted three days but killed no one.

After the telephone call, Newhall pulled some topographic maps of the Pinatubo region that he had in his office. Not much was known about Pinatubo, and that was always a concern. "The ones that have been sleeping the longest and are the poorest known are potentially the most serious ones," says Newhall. He did a search of the literature and found that a geothermal study of the region had been done in the early eighties. Newhall tracked down the geologist who had mapped the region for the project. He was then studying in Florida and he sent Newhall more maps.

The maps showed that Pinatubo wasn't an impressive mountain. In fact, it was probably a dome that had formed inside a large caldera. The most interesting feature Newhall saw in the maps was an unusual drainage system that surrounded the mountain. The pattern was known as dendritic drainage because it was a complex jumble of small streams. A pattern like that is often a signature of big eruptions. When pyroclastic flows hundreds of feet deep come to rest, rains easily slice the loose ash, eventually creating canyons, and the pattern of water runoff begins to look like the branching of a tree. This pattern told Newhall that Pinatubo, which had no historical record of erupting, had left footprints of at least one large eruption. The size of the patterned ground suggested an eruption as large as Krakatau's.

With a little more research, Newhall found that about a million people lived on the gently sloping and highly fertile apron of the volcano.[1] In fact, *pinatubo* means "to grow."[2] It was likely that the rich farmlands were produced by past eruptions. The current crowding around Pinatubo meant that if it erupted on the scale of Krakatau, the blast would overwhelm nearby Angeles City with its three hundred thousand citizens. It would also reach Clark Air Base, the largest U.S. military base outside the United States, which was fifteen miles from the summit.[3] And it could make life uncomfortable at another U.S. military installation, Subic Bay Naval Air Station, twenty-four miles south of the summit.[4]

The current activity probably wasn't caused by rising magma, Newhall thought. Less than a year earlier, in July of 1990, the island had been slammed by a 7.8 earthquake, which brought down new hotels and old schoolhouses and killed sixteen hundred people. Two weeks after the quake, a group of nuns had showed up at PHIVOLCS and reported that the local volcano was shaking and producing new steam vents. Pinatubo wouldn't be the first volcano to be reawakened by a tectonic earthquake. But a quick check by a PHIVOLCS team found nothing that raised serious concern. The most likely explanation was that Pinatubo was already a warm volcano, which is why geothermal explorers had been interested in it. The quake had probably rearranged some of the volcano's plumbing, exposing some pocket of underground water to a new conduit of heat. The result would be new steam vents.

When the same group of nuns returned a year later with reports of new unrest, Punongbayan dispatched another team. They began detecting plenty of small earthquakes centered five miles northwest of Pinatubo's summit. By mid-April, the area near Pinatubo was fluttering with 30 to 180 quakes a day. Punongbayan ordered an evacuation from within a circular zone six miles from the summit.

On April 16, Newhall finished his background research. He wrote a six-page summary and faxed it to Punongbayan. In it, he said the unrest at Pinatubo was "particularly worrisome and particularly difficult to interpret." It was worrisome because of the large number of people potentially at risk, and it was difficult to interpret because of the limited monitoring that had been done. He said the possibilities ranged from a purely hydrothermal event to the ascent of magma. "If Pinatubo were to erupt, the eruption would probably be explosive and serious," he wrote.

He advised that pilots, both civilian and military, should be warned that volcanic ash can melt jet turbines and fuel nozzles causing planes to lose power. Finally, he mentioned that the Survey's VDAP team might be able to offer some assistance "provided that external funding can be found."

Newhall e-mailed the VDAP group at the Cascade Volcano Observatory asking them what they might be able to do on short notice. They replied that they could put together a portable seismic network, some tiltmeters, and maybe a COSPEC. But they had no money. Neither VDAP nor the Survey had an extra $100,000 to buy a new cache of equipment. In fact, at the time the Survey didn't even have money to buy airplane tickets to the Philippines for a VDAP team. The equipment in VDAP's possession actually belonged to US AID and had been purchased for use on volcanoes in Central and South America. VDAP had never responded to a volcano outside of the Americas, and some Survey officials thought it had no larger mandate. Finally, the VDAP team reminded Newhall that, if they were going to work in another country, they needed to be formally invited by the "responsible agency."

Newhall passed that information on to Punongbayan, who soon submitted a formal request for VDAP assistance. Then Newhall attempted to get the State Department's approval to respond to the Philippines. But its Office of Foreign Disaster Assistance, which funded VDAP, was preoccupied with problems growing out of the recently completed Gulf War. They had their hands full coordinating humanitarian relief to the Kurds in northern Iraq. The few resources that were left over were directed to Bangladesh, where a spring flood had just devastated the country's large and heavily populated delta region.

Newhall phoned Manila. He was looking for help from the US AID office there. The call was not warmly received. American diplomats had their own problems, many stemming from the previous year's earthquake in Luzon, which had killed several of the embassy's own people when a hotel collapsed on them during a conference of private international assistance groups. The US AID office didn't have an extra dime, they said impatiently.

Another factor that influenced the US AID response to Newhall was that the United States was renegotiating the leases for the Clark and Subic Bay military installations. The forty-four-year-old lease the United States had with the Philippine government for Clark was set to run out on September 16, 1991. The United States had proposed a ten-year extension at $360 million, but the Philippine government wanted $400 million.[5] The ensuing bargaining had been front-page news for weeks and it had inflamed passions. Many Filipinos did not want the American military to remain for any amount of money. Tensions were running high in the country, a few Americans had even been assassinated, and US AID was under pressure to reduce the number of American personnel in the Philippines, not bring in more.

When the US AID office in the Philippines turned down Newhall's request, Newhall phoned the chief air force weatherman at Clark. "I figured he'd be a kindred spirit, a fellow scientist, and he was," recalled Newhall. He quickly learned the weatherman had been a graduate student at the University of Washington when Mount St. Helens had erupted in 1980. Newhall explained that he was calling because he had received a report from PHIVOLCS about unrest at Pinatubo. Had anyone on the base noted anything unusual? Yes, the weatherman replied. Substantial columns of steam had been visible from the base for the last two weeks. Also, residents in the portion of the base closest to the mountain, an area of family housing known as The Hill, had complained of unusual odors. A team trained to detect poison gases had taken samples but reported that the pungent odor wasn't harmful.

As Newhall had hoped, a few days after his conversation with the air force weather officer, Newhall received a call from Colonel Bruce Freeman, the number two ranking officer at the base. Freeman was concerned about the steaming. He had essentially one question: "When can you get here?" Newhall told Freeman no funding was available from the Survey, State's disaster assistance office, or US AID in Manila.

Three days later, Newhall received a terse fax from US AID in Manila that read: "This Mission asks that you come to the Philippines to help us with Mount Pinatubo. . . . We have a maximum of U.S. $20,000 for this effort, do not feel that [you] have to spend it all." Newhall chuckled. He was about to deploy $100,000 worth of equipment that would probably never be returned, so before he could even get a plane ticket, he was $80,000 in the red. But he decided to worry about that later and called the VDAP offices to give them the green light.

When putting together a team to respond to any volcano at that time, the first choice on anyone's list would have been Dave Harlow. Harlow was known as the best intuitive volcano seismologist on the

planet. Harlow had monitored more than fifteen volcanoes since starting with the Survey, most of them from their earliest quiverings to their
final eruption. But in the university-like culture of the Survey, Harlow
was a "meatball," a person who may have been good in a volcanic crisis
but someone who had produced few substantial publications. Career
advancement and pay were directly connected to a scientist's record of
publication, rather than to how many lives he may have saved. In fact,
Harlow had been so busy running off to restless volcanoes rather than
publishing research papers, he had been stuck in the same government
pay grade for thirteen years. Now, he was working for one of the Survey's most respected seismologists (a type of scientist the meatballs
called "conehead"), who advised Harlow he would do his career little
good by going to Pinatubo. It was Harlow's last chance to salvage his
career. So Harlow said thanks, but he had to pass.

Newhall next called John Power, who was then living in a sixteen-by-
sixteen-foot cabin (with no running water, and a woodstove for heat)
ten miles outside of Anchorage. Power was a young but solid Survey
seismologist who had helped run the response to Redoubt volcano in
1989. Newhall's call came just as Power was digging a path from his
home to the road through a six-foot snowbank. The temperature was
twenty degrees below zero when Newhall asked him if he wanted to go
to the tropics. Are you kidding? Trying to save money, Newhall asked
Power if the two of them could manage deploying a portable volcano
observatory and monitoring the progress. Power said he was comfortable with the software and data analysis of the seismic network, but
they would need a more experienced hand to install and maintain the
instruments.

So Newhall tapped one of VDAP's top hands at CVO, Andy Lockhart,
a geophysicist who had prospected for gold in Alaska before joining the
Survey. Lockhart was a bright, funny guy who was a wire-twisting wizard, but he had a habit of getting sick during a volcano crisis. Often, a
chronic sinus infection would flare, and occasionally he would come
down with something more exotic, such as typhoid, which he picked up
in Guatemala.

The final member of the team was Newhall himself, who would handle the COSPEC and do the hazard mapping. If something was going to
happen in the Philippines, Newhall would not miss it. He passed off his
aircraft/ash conference duties to a colleague and, on April 22, hopped a

plane to Seattle. There, he met Power, Lockhart, and their thirty-five trunks of equipment. Over the Pacific, they outlined the work ahead of them. As they saw it, the job was to help PHIVOLCS get some additional instrumentation on Pinatubo to clarify what the volcano was up to. Newhall was aware that this wasn't exactly what the Air Force had in mind. The USAF thought the geologists were coming to provide them with guidance. Newhall thought the different goals would mostly overlap and if they didn't, they could be worked out later.

The volcano team arrived in Manila in the early-morning hours of April 23. The temperature at 4 A.M., Power noted, was ninety-two degrees. They were met by staff from the U.S. embassy, who had brought several bulletproofed Chevy Suburbans to transport the equipment and the scientists.

The official greeting later that morning from US AID officials was brief and cold. The AID mission in the Philippines was overworked trying to provide food, housing, and medicine. They were under pressure to accomplish their mission with fewer and fewer people, as demanded by the Philippine government. The mission was also still reeling after losing staff members in the earthquake just months before. Now, as they looked at the scientists sitting in their offices, they were not happy about a phantom crisis—a puny volcano that might never erupt. As they saw it, this additional burden was being thrust on them by a bunch of scientists eager to do some research in an exotic location.

After the brief greeting, the geologists piled back into the Suburbans, and the caravan of vehicles with diplomatic plates snaked its way to Manila's neighboring Quezon City, to the PHIVOLCS headquarters. The agency's crowded offices were on the upper two floors of a building that seemed to be decaying in the heat and humidity. Inside, the noise of the street jeepneys was overwhelmed by the rattling of the air conditioners. The three-member VDAP team was escorted into Punongbayan's office, which was crowded with young Filipino geologists and technicians. After the customary Nescafé was served, the latest data was spread out on Punongbayan's desk.

The seismic data was the most interesting. Power looked carefully for any traces of moving magma that might be hidden in the noise of larger earthquakes. He found none. The PHIVOLCS team spread out a map and pointed to a spot three miles northwest of the steam vents. This, they said, was the origin of the quakes. Both Power and Newhall doubted it. The quakes were right under PHIVOLCS seismic net, not under the vents where they should have been. It didn't make sense that this hot system that was creating vigorous fumaroles was being heated by a plug of magma three miles away. More than likely the Filipino scientists had been misled by the limited number of seismic stations they had. They were also limited in the locations they could use. Because the PHIVOLCS team did not have helicopters, they were restricted to areas they could reach by roads. And Pinatubo was a largely roadless, dense jungle broken by steep canyons. So the PHIVOLCS instruments were clustered on the northwest side of the volcano, which coincidentally was precisely where they thought the earthquakes were originating.

The most visible sign of trouble came from the three steam vents that lay along a mile-long canyon on the upper north flank of the volcano, at the heads of the Maraunot and O'Donnell River valleys. Each vent was large, about the size of a baseball field. And each was jetting such an impressive amount of steam that they were easily seen twenty miles away.

The two groups of volcanologists mapped out a plan. There were three jobs to do. First, they had to step up the monitoring. That meant putting more seismometers on the volcano and beginning to take COSPEC measurements for gas emissions. Second, they needed to literally start digging up the history of Pinatubo's eruptions. Just as Crandell and Mullineaux had done at St. Helens, the geologists had to determine if the eruptions had any periodicity and identify which areas would be threatened in a new eruption. Third, after they knew Pinatubo's past and present better, they had to educate provincial officials, the Catholic nuns, the village chiefs, and the Air Force generals about the consequences of an eruption.

The team then drove sixty miles to Clark through rice paddies and jungle that looked a lot like Vietnam war movies. (It was no coincidence. Many war films, including *Apocalypse Now*, were filmed there.) Approaching the base, the geologists spotted a spike on the horizon, a cone-shaped volcano that loomed above the surrounding green countryside. That was Mount Arayat, which stood 2,700 feet above the surrounding landscape. Pinatubo was a bump in the distance, only 650 feet above nearby hills. It, too, was easy to spot. Just to the north of the summit, substantially down from the peak, steam vents were billowing a white column into the clear blue sky. The fumaroles looked like smoke signals.

Then they drove through the main gate at Clark Air Base and into another world.

Clark occupied a special place in American military history.[6] It had been a U.S. outpost since the Spanish-American War. Early cavalry planners were drawn to the location by the grass, which was better for the horses than that of other grazing areas on the island. Then, the Second World War had made Clark sacred ground. Clark had been nearly destroyed by a Japanese attack on December 8, 1941. Within two weeks, the base had been evacuated in the face of an overwhelming Japanese force. After holding out for months on the nearby Bataan peninsula, more than seventy thousand American and Filipino soldiers surrendered to the Japanese in April and were marched through the jungle to an internment camp. Along the way, they were starved and beaten, and many were run through with bayonets. Only fifty-four thousand completed the fifty-five-mile walk to reach Camp O'Donnell. The Bataan Death March, as it came to be known, went right through Clark. Not until 1945, when General Douglas MacArthur returned to the Philippines, was Clark reclaimed. A promise was made then that Clark would never again be abandoned.

In more modern times, Clark had become an important way station for troops and aircraft during the Vietnam War. By 1991, Clark had become uniquely important to the U.S. military as a stage for aerial war games. The mock battles were fought in the skies over nearby Crow Valley. Nearly every day of the year, pilots from the Air Force, Navy, Marines, and foreign allies fought their way through sham SAM attacks and enemy fighters to bomb the wicker airports, bunkers, and infrastructure in Crow Valley. The stage was an invaluable resource for the United States. Because of it, in 1991 Clark was growing with $150 million in construction contracts for that year alone.

The base itself was roughly triangular with its tip pointing west at Pinatubo. At the other end, its runways lay inside the triangle's broad base on the east. The highest point on the base was at the western tip, a neighborhood called The Hill. This family-housing area had nearly two thousand homes, schools, and stores. From the tip of The Hill, the base gently sloped away from Pinatubo to the runways. At its heart was the parade ground, which was encircled by enormous acacia trees, some standing eighty feet high. Around the parade field there were offices, a golf course, and more housing, and the entire base was encircled by an eight-foot-high concrete wall that was twenty-six miles long. From the mountains to the west, two rivers flowed down the sides of Clark's triangle: the Sacobia to the north and the Abacan to the south. Each river had lesser streams that cut through the base.

Despite the American-Filipino bonds forged in World War II, Clark and nearby Subic Bay Naval Air Station did not rest comfortably within the modern Philippines. A strong anti-American sentiment had arisen, and passions had intensified during the ongoing base negotiations. The anti-American passions helped fuel a guerrilla movement that operated in the jungles between the base and Pinatubo. And recently several Americans had been assassinated near the base. One victim had been exploring the geothermal resources around Pinatubo when he was gunned down on a back road. Another was a young soldier who was walking just outside the wall of the base when he was shot in the back of the head.

The protests and murders strained the already tense relationship between Americans and Filipinos on Clark. While many families living on the base employed live-in maids (for a salary of $6 a week) and gardeners, the prevailing view was that most Filipinos were murderers or thieves. Theft was such a problem on the base, from people who "came over the wall," that the running joke was that "everything on the base is stolen, it just hasn't been picked up yet."

Newhall ran into this bias at the main gate to Clark where he found Punongbayan waiting. The PHIVOLCS chief was not allowed to enter. Filipinos were generally denied entry to the base without an official reason and proper credentials. Newhall, whose sensitivities to the Filipino culture and people ran deep, vowed right then that he would not locate the VDAP observatory on the base. He worried that his Filipino colleagues would be subjected to harassment every time they needed to pass through the gate. Moreover, if the crunch came and warnings of eruptions needed to be issued, Newhall wanted those warnings to be free of the appearance of any political motive. Without such credibility, people could die.

The main gate at Clark was a portal between cultures. On one side was Angeles City, a teeming, dirty, vibrant Third World urban center with a zone next to the base of sex bars, trinket shops, and food stalls. On the other side of the gate was suburban America with its own Baskin-Robbins and Pizza Hut. The base's television station even broadcast American TV shows, including news and sports.

"It was the damnedest thing," geophysicist Andy Lockhart recalled. "It looked like America as you would imagine it, if you were designing it from memory and glossing over the rough spots. The place is surrounded with this huge expanse of grass. No litter on the place at all. All the streets are clean, everybody is clean-cut. It was kind of creepy. You kept looking for the bad part of town and not finding it. It was like this Heritage USA theme park which was like the U.S. but not the U.S. And right outside the gate, pressed up against this mom-and-apple-pie world was the most corrosive layer of lechery and filth you can imagine with this zone of bars, the Buster Hymen's and the Nepa Hut, where the live sex was just the warm-up act.

"Chris found this really distressing and he was hugely embarrassed by it. He hated being on the base with the PHIVOLCS staffers, in there where the guys were taking advantage of the poor local women who were in turn prostituting themselves to these guys. Chris loves the Philippines and he loves Filipinos, and to see that kind of ill usage brutally offended him."

The first three days after they arrived, Newhall and his team hunted for a place outside the base for their observatory. They looked at schools, hotels, and even a church. Nothing was right. They needed to see the volcano from the observatory, which was difficult from most locations because Pinatubo sat so low on the horizon. They also needed a spot that was free of stray radio signals so they could get good data streams from the instruments. And in an ideal world, they needed a reliable source of electricity, which would be especially important if the infrastructure was collapsing around them during an eruption.

Most of all, they needed helicopters to get around Pinatubo. The volcano had only one good road, and that was on the west side, away from the base and the fumaroles and the earthquake cluster. A drive from Clark to the west side and back could take all day.

Shortly after they arrived, Punongbayan arranged for a flight in a Philippines army helicopter. Andy Lockhart took the flight to search for spots to plant seismic stations. The first time the military helicopter swooped low to land, the door machine-gunners readied their weapons and scanned for any sign of guerrillas. Andy's heart sank. He imagined that every time he needed to go out and change batteries on a station, it would be a military assault.

At Clark, the Air Force officers became concerned when they learned that Newhall was looking off base for his observatory. After three days of searching, just as Newhall was coming to believe that there was no good site elsewhere, Colonel Freeman asked Newhall what he would need from the Air Force to work on Clark. Newhall negotiated for communications, living accommodations, and helicopter support. He also wanted the PHIVOLCS staff to live on base in the same quarters offered to the VDAP team. He wanted them provided with gate passes and a promise they wouldn't be harassed by the gate guards. Agreed.

Like the Forest Service at St. Helens, the Air Force wanted the geologists on the base for lots of reasons. First, it gave them ready access to any important information the scientists might develop. But it also gave them more control over the scientists. A few of the senior officers had been worrying that one day an irresponsible geologist would call a press conference and tell Filipino reporters that the entire region should be evacuated, including those fools inside the base. If it ever came to a decision to evacuate, the decision would be made by the Air Force. It would not be the result of pressure put on them by a bunch of "beards" wearing shorts and flip-flops asking reporters, "Why are those people still on that base?"

"I was impressed with these [Air Force] guys," said Lockhart, who had had plenty of experience working volcano crises with local mayors and governors in Latin America. "They were trained decision makers. It was not like working with elected officials that you meet elsewhere, and the first time you meet them you look them in the eye and you want to say, 'I am the last guy you want to see walking through your door because the shit is about to happen, man, and you have no idea how bad it's going to be.' You feel sorry for these guys because you know you're going to be their fucking nightmare. But these [Air Force] guys, these guys were ready to deal with that."

The geologists inspected several locations on the base for their mini-observatory, but on Clark, too, the low-slung volcano could be seen from only a few spots. Eventually, they found a three-bedroom town house among enlisted men's housing on Maryland Street in the center of the base. The apartment was a surprisingly good location. The second-story bedroom window offered the geologists a view of Pinatubo over the houses across the street.

Within hours of their settling on the Maryland Street apartment, a truck pulled up outside and servicemen began carrying in and unpacking the "condo kit." Out of boxes came beds, linens, towels, sheets, dishes, wineglasses, pots, alarm clocks, a television, and a microwave. Five hours after arriving, the condo-kit people were gone. VDAP folks, who were used to much more primitive conditions in Latin America, began to think working with the Air Force wouldn't be so bad after all.

The VDAP team began unloading the "observatory kit" from the thirty-five trunks they had brought. Most of the receivers, computers, and drum recorders went into the small front bedroom on the second floor. This room became the data center. Desks, radio receivers, a fax, phones, maps, scientific papers, books, and seismic drums soon filled the room. One drum recorder was placed just below the window, so a scientist could watch the earthquake activity being scratched out and look up at the mountain to see if it was related to any change in the volcano. Outside the town house, the VDAP team planted antenna masts to receive the radio signals from the monitoring equipment. Eventually, Newhall erected a sign out front that read, "Pinatubo Volcano Observatory—Shake and bake with the best."

The priorities were to do a quick geological reconnaissance to identify danger areas and determine if the eruptions had a periodicity, and to plant the seismic network so that locations could be better plotted and magmatic signals could be detected if they were present.

After watching the condo kit transform the bare apartment into a home in an afternoon, Andy Lockhart thought the resources of the Air Force were boundless. And what Lockhart wanted most was a helicopter. The only way to lay a seismic net wide enough to get good coverage was by going in by helicopter. Without broad coverage, VDAP, too, might be vulnerable to misidentifying the location of the quakes. The Philippine military balked at providing a helicopter for Americans working from Clark, so Andy was left with the USAF.

The base had dozens of high-performance fighter jets such as the F-4E Phantom and F-4G Phantom II Wild Weasels, but it had only three helicopters. By a quirk of modern military custom, planes belonged to the Air Force and helicopters were owned by the Army. There were a few exceptions, which is why Clark Air Base operated three Hueys. Peri-

A pyroclastic flow with various other volcanic material embedded was found near the officers' quarters at Clark Air Base in the Philippines by Chris Newhall soon after his arrival. (PHOTO: CHRIS NEWHALL)

odic maintenance regularly put one of the three out of commission, and demands for others—to fly senior officers to Subic or to Manila for base negotiations—meant the Survey scientists were at the bottom of the list.

Not waiting for a chopper, Newhall, Power, and a few PHIVOLCS geologists drove in the embassy-provided Suburban into the field to begin digging up the details of Pinatubo's past eruptions. Near the edge of Clark, Newhall stopped in a canyon with walls 230 feet high. He got out of the vehicle and, in complete awe, gazed at the walls, which were not rocks but layer upon layer, maybe a dozen in all, of lahar and pyroclastic flow deposits.

"Hooooly moooley," Newhall said to Power. "Look at the size of these deposits."

Newhall's preliminary assessment, which he had formed while back in Reston, was that when Pinatubo erupted, it was big. But he wasn't expecting it to be big enough to produce these deposits. A few pyroclastic flow deposits were thirty feet thick. He estimated the spot he was standing on was almost ten miles from the volcano, and to see thick pyroclastic flows at this distance was worrisome. At St. Helens, pyroclastic flows rarely got farther than a few miles from the vent, and those deposits were thin. This evidence of Pinatubo's past performance meant the volcano had a history of enormous eruptions. These eruptions had put big pyroclastic flows into the river that flowed along the northern wall of the base. The record staring Newhall in the face also suggested that these same events had probably swept across what was now the Clark Air Base, incinerating everything in their path.

Newhall rooted around in the wall of the canyon searching for bits of charcoal. Finding one, he would package it and map its location. These samples were flown back to the Survey's radiocarbon lab in Reston, Virginia, where they were dated. Within days, the results were faxed to the Pinatubo Volcano Observatory. Pinatubo had erupted about every thousand years, give or take five hundred years.[7] The volcano had last erupted six hundred years earlier, so it was possible the volcano was now cocked.

While Newhall was making headway, Lockhart and Power were regularly floundering at the terminal where a snarly officer told the scraggly geologists the helicopters were needed elsewhere that day. Come back tomorrow.

To kill time and check out the equipment, the two men put in a seismic station in the middle of the base. On April 25, they planted the instrument on a spot known as Lilly Hill, where the Japanese had made their final stand in World War II. The geologists designated the site LiL, for Lilly Hill, on their maps. The sensor, the transmitter, and the receiver worked just fine, but the data was garbled with seismic noise. Returning to the site, they discovered a water tank one hundred yards from the instrument. Whenever water was pumped into the tank, the ground quivered slightly, just enough to make the seismometer unreliable. So they relocated LiL a mile away, on another hill, renaming it CAB—for Clark Air Base. It was only marginally better, and eventually the CAB instrument's battery died and the instrument was nearly forgotten.

When the Philippine heat, the surreal surroundings, and the continuing frustrations with the helicopter officer got the better of them, Power would turn to Lockhart and say, "Andy, what do you want to do today?"

"Well, John, let's go to the Philippines." They would motor through the gate and around Angeles City, and often enough Lockhart would say, "This is lovely, in the Philippines."

While the team had brought much of the equipment needed, they did not bring batteries for the seismometers. They were difficult to transport because airlines didn't like to carry acid in the baggage holds, and generally batteries were available in the host country. Newhall found the batteries the team would need at the base exchange, a government-run superstore for base personnel. But he didn't have enough money to buy the dozen batteries, so he floated them on his personal credit card. The VDAP folks often spent their advances quickly, then financed the rest of the operation on their credit cards, sometimes to the tune of thousands of dollars. Upon returning home, they also often found pointed reminders that they had exceeded their spending limits. One geologist returned from a Central American crisis with his credit cards maxed out in what was essentially an interest-free loan to the Survey only to find the Survey had had his wages garnisheed.

Eventually Lockhart and Power got more priority and helicopters became available. They began flying at every opportunity. At first, they were looking to appropriately distribute the seismometers. They knew they would get the best data if the instruments were well separated from each other, blanketing Pinatubo in every direction. Plotting the best locations on a map in Maryland Street was easy, but flying over the country around Pinatubo was a lesson in radical geology. In all directions, Pinatubo had pumped out broad sheets of pyroclastic flows that had compressed into broad plains of ignimbrite. Water had cut steep canyons, sometimes hundreds of feet deep, into these ignimbrite sheets. The fingers of flatland between the canyons were overgrown with dense forests and tall elephant grass. It was an extreme topography of sharp peaks and valleys, all coated with dense vegetation.

On April 30, the team put in its first real field seismometer on a ridge two-thirds of the way from the base to Pinatubo. It was high enough for a line-of-sight radio signal to Clark. The ridge was one of two running due east from the volcano. Looking up the valley between the two ridges was like looking down an alley from Pinatubo to Clark.

The ridge was covered with twelve-foot-high elephant grass. Lockhart told the pilot to hover near the ground, so the prop wash would plaster down the grass. Then Lockhart jumped out with two PHIVOLCS technicians and a few bags of equipment. As soon as the helicopter moved away, the grass stood back up, and the geologists began slicing it away, using shovels as scythes. Power headed back to Clark in the chopper, landed, and drove outside the gate to the nearest tourist stall where he bought souvenir bush knives, which he took back to the team working on the ridge. Soon, John Power, who just a few weeks earlier had been shoveling snow in Alaska, was hacking away at elephant grass in temperatures well over one hundred degrees Fahrenheit, with humidity soaring.

Planting the station was exhausting work. When the ledge was cleared, the crew began digging a hole big enough to bury a garbage can. Field monitoring equipment is generally buried to keep it from being influenced by weather and to keep it out of sight of vandals. After the hole was dug, they dug a trench for the cables that ran from the can to the radio mast. They dug a hole, prepared and poured concrete, and set the mast.

With the holes dug, they planted the working end of the seismometer. The geophone was about the size of a Foster's beer can. It is essentially a mini-generator. Inside the instrument is a magnet on a spring that sits inside a coil like a missile in a silo. When the earth moves, the seismometer case and coil both move with it. The magnet tends to stay in one place, due to its inertia. So the coil moves around the magnet, generating a minute pulse of electricity. The voltage is amplified and converted, by a microchip, into a tone. The tone is radioed to the seismograph, which converts this sound of a quake into the motion of a pen or a dot on a computer screen. When mountains are really bouncing, some volcanologists say the volcano is singing.

The geologist digs a narrow hole for the geophone as deep as he can reach into the ground. Then he pours some quick setting cement into the bottom of the hole and sets the geophone into it, lying on his stomach and holding the geophone level with a bubble level as the cement sets up. The instrument is wired, then the wires are run through the ditch and attached to the mast. The mast has four holes drilled into one end, through which the geologists hammer lengths of rebar. This acts as footing and holds the mast erect under everything but the most extreme conditions. Finally, the site is named and mapped. This one was known as PIE, for *PI*natubo *E*ast.

After PIE, the instruments started going in quickly. UBO (named because on a map it was located in the center of the *o* in Pinat*u*bo) went in a half mile east of the summit on May 1. It was on the site of a drilling pad from an abandoned geothermal project. BUR was planted northwest of the summit and its signal was relayed through the Air Force outpost at a place called Bunker Hill. PPO was set on a concrete floor inside a cinder-block guardhouse on May 3, a mile north of the summit and a mile from the vents, near a native Aeta village named *Patal Pinto*. PPO was close to where the PHIVOLCS geologists had located the source of the earthquakes. BUGZ went in about four miles from the summit. Because BUGZ was located on the southwest side of the mountain, its signal also had to be relayed to Clark. The installation of an instrument and a radio repeater at Negron, halfway between PIE on the east and BUGZ on the southwest, completed the net on May 14.

The pattern was roughly circular. PIE was at three o'clock; Negron at five; BUGZ at seven; BUR at ten; and UBO, nearly dead center. CAB was the most remote instrument from the mountain, on Clark, which gave it the potential of filtering out smaller events if Pinatubo heated up.

Four of the instruments on the east side of Pinatubo were transmitting radio signals by line of sight directly back to Clark and were picked up by the mast planted outside Pinatubo Volcano Observatory (PVO) on Maryland Street. But two stations on the north side of Pinatubo were transmitting to an Air Force outpost in the Crow Valley bombing range, called Bunker Hill, where their signals were relayed by phone lines to PVO. BUGZ was also relayed, through Mount Negron.

With the net complete, the VDAP/PHIVOLCS team began pinpointing the location of the earthquakes. The primitive system established by the PHIVOLCS team on the west side of the volcano was still placing the earthquakes about three miles northwest of the summit. The PHIVOLCS field team was encamped in a schoolhouse and used three-day recorders, which produced a floppy-disk record that had to be physically retrieved and measured by hand. Then they had to wrestle with the calculations. All the locations they were reporting were right underneath their net. The VDAP team just didn't believe it.

Even before the net was finished, the VDAP team was confident they had enough data to fix the location of the quakes. It was the same spot the PHIVOLCS team had identified, three miles northwest of the summit. The confirmation didn't clarify the situation. In fact, as Power began pinpointing each new quake, the mystery over the quakes grew.

The earthquakes were clustered three miles northwest and three miles down, in a region that Power and Lockhart began calling "the northwest cloud." The geometry of the cloud didn't look volcanic. It wasn't organized in any manner, such as under the fumaroles, nor were the earthquakes migrating toward a vent as seen at Hawaiian volcanoes. But they also didn't have any linearity to them that might indicate that they were occurring along a tectonic fault line. The Survey geologists just didn't know what to make of it.

The seismic data didn't shorten the list of possibilities. The 7.8 earthquake the year before could have released some heat source beneath the volcano that fueled the fumaroles. And the northwest cloud could be unrelated to the fumaroles. Theoretically, Pinatubo was within what seismologists call the "area of influence," and the cloud was close to a known fault. But volcanic activity at such a distance from the volcano would be rare. The scientists began creating scenarios to account for distance between the quake center and the fumaroles. Perhaps displacement along a distant tectonic fault had disturbed the volcano's hydrothermal system, which then led to the fumaroles. But essentially they were stumped.

Toward the second week in May, a disjointed picture of the volcano was forming during the evening science meetings, which were held in the living room at the Maryland Street condo.

Pinatubo was a volcano that, according to the carbon dating of its eruptions, might now be primed to reawaken. When it had erupted in the past, the eruptions were huge, with pyroclastic flows that would cover today's villages and towns and would at least rush up to the gates of Clark Air Base.

But it was not clear that the current activity was driven by moving magma. Vigorous earthquakes and intense fumaroles were indications of magmatic activity, but they could also be disruptions, caused by a recent large earthquake, of the volcano's hot-water system.

Newhall hoped to use the COSPEC to confirm whether magma was driving the activity beneath Pinatubo. The COSPEC had been of no help in predicting the May 18 eruption of Mount St. Helens. But on several volcanoes in the 1980s, the instrument did detect sulfur dioxide (SO_2) escaping. Then a CVO scientist named Terry Gerlach discovered that some volcanoes, "wet" volcanoes he called them, such as St. Helens, are so laden with water that SO_2 escaping from the magma is absorbed in the water. It doesn't freely reach the surface. So Newhall knew that if the COSPEC detected high levels of SO_2 over Pinatubo, it would increase the likelihood that magma was present, but if he found nothing, it wouldn't destroy the idea either.

"We desperately needed some hard data from the COSPEC to compare against the seismic data," said Newhall, "to be able to say, 'Look, this is not just an aftereffect of the earthquake last year. This volcano is stirring.'"

Newhall felt desperate for several reasons, all grounded in the skepticism VDAP was encountering. Outside Clark, Newhall had run smack into it when he and Punongbayan had attempted to brief government officials in the region. In fact, the mayor of the largest city threatened by the volcano, Angeles City, refused to even meet with Newhall and Punongbayan. He told the local papers that the worry about an eruption at Pinatubo was an overreaction.

Things were not much better inside Clark. The base commander, Colonel Jeffrey Grime, said he had a better chance of winning the California lottery than of having the volcano erupt. His skepticism was generally, but not universally, reflected in his subordinate officers. The VDAP team believed that if Colonel Grime took the threat of an eruption seriously, then they would have a higher-priority claim on the limited helicopters. Newhall also wanted to begin a broad education campaign, including appearances on the base TV and a series of stories in the newspaper, aimed at those people living on the base and nearby, about the many hazards a volcano presented. He had already been tutoring the senior staff (although not Grime) in volcanology and volcano monitoring. He had the base television station make copies of the Krafft's hazard video, and he passed those out at briefings, which the officers began calling Geology 101. But Grime refused to allow a broader education campaign.

Grime also opposed suggestions for evacuation planning. Some of Grime's senior officers had requested permission to begin the job, just in case things turned bad fast. Although Grime shot that proposal down quickly, one of his officers, Colonel John Murphy, began to do it anyway. Murphy enlisted another officer in the plan, and the two began scouting for fallback positions. Murphy envisioned a worst case and concluded the base would have to be completely evacuated, not just The Hill residential neighborhood, but the next five miles down from The Hill to the other side of the base. Murphy found a fallback position in the shadow of Mount Arayat, at the Pampanga Agricultural College ten miles east of the base. Arayat was twenty-seven hundred feet higher than anything else around. In fact, US AID officers, giving tours of the potential trouble spot, often mistakenly pointed at Arayat as the currently threatening volcano. The college officials agreed to let the Air Force store food and water there in exchange for some improvements to the buildings.

Newhall also was under pressure from the Survey itself. Back at the Survey field office in Menlo Park and at CVO in Washington State, senior Survey officials had begun to panic about financing the Philippine adventure. They called Newhall frequently, usually late at night or early in the morning because of the time difference. Each call was a reminder that Newhall had taken the only mobile volcano-monitoring equipment the Survey owned. Technically, the equipment Newhall took belonged to the State Department for use in the Americas. What would happen, they asked, if a volcano crisis flared up in Central America now? They had no extra equipment. Moreover, Newhall's gang was spending money the Survey didn't have. How could that be repaid? The only source of funds would be to curtail the field seasons of other Survey geologists, work that had been planned often years in advance. How could they justify that for a volcano with a slight fever? And Pinatubo could steam and rattle indefinitely. It could consume money for months without erupting.

Each call got Newhall steamed, and he began to let it show. Like the geologists who responded to St. Helens, Newhall wasn't pacing himself. There was too much to do, and it all was important. People's lives could depend on the work he and his team were doing. Now these incessant calls not only disrupted the little sleep he was able to grab, but they magnified the problems of heat, racism, paranoia, and skepticism that clung to this crisis response like the midday humidity. At one point, he snapped and said, "We've got enough things to worry about over here with the volcano. The last thing we need is to be having someone undercutting us back home."

So Newhall was hoping for a lot from his COSPEC reading, mostly support. It was the one thing that could confirm that fresh magma was driving the unrest. Or it could show nothing, which wouldn't mean a thing. A negative COSPEC reading might mean that Pinatubo was like

St. Helens, one of those explosive volcanoes that didn't produce a COSPEC signature.

To get a good COSPEC reading, Newhall needed an airplane. But, as with the helicopters, the U.S. Air Force couldn't help much. Clark was overflowing with all types of high-performance jets, but it was lacking in what Newhall really needed—a small, slow, propeller-driven plane. Clark's flying club had such a plane, but it had been grounded following a recent accident. Newhall nearly secured a light plane from the Navy's Subic Bay, but when he told the pilot that he wanted to fly low over the volcano that was blasting the steam jets, the pilot backed off.

By the second week in May, Newhall figured out a way of rigging the COSPEC on one of the helicopters. Generally, to do a COSPEC run, the instrument needs to be flying under the invisible gas plume and pointed straight up into a clear sky. The difference in the ultraviolet spectrum between what the instrument sees and what it should see is translated into the amount of SO_2 in the air. Small planes are best because they can fly around the volcano in minutes and have windows that allow the instrument's telescope to be pointed vertically. When Newhall rigged the instrument on the Air Force helicopter, it was pointing at a forty-five-degree angle to miss the helicopter blades. Though a creative solution, it might not provide usable or, more important, reliable data.

On the morning of May 13, Newhall lashed the sixty-pound instrument, which looks something like a tank turret, to the gunner's door of a helicopter. He instructed the pilot to fly in circles, low over Pinatubo. On several passes, Newhall could see the needles rise and fall. When he returned to Maryland Street, he worked through a long series of calculations to get what volcanologists call the concentration profile. Pinatubo, he found, was releasing five hundred tons a day of SO_2 into the atmosphere.

The figure was actually not huge, but to Newhall it confirmed the presence of magma. This wasn't a hydrothermal system that had been knocked out of whack by an earthquake. It surely wasn't activity from a tectonic fault. This event was driven by a tongue of red-hot magma. And it was close enough to the surface to be degassing.

It was impossible to say how much magma was involved or how close to the surface it was. And it was impossible to predict if the

View of the north flank of Mount Pinatubo from a helicopter in late April. Note active fumaroles and dead trees from the blast on April 2.

magma was rising and would lead to an eruption. But the probability of Pinatubo erupting at all jumped with the detection of SO_2.

What was the probability, Newhall wondered, of an eruption? To answer that question, he began creating a probability tree. At about 3 A.M. on May 17, he sat down in the drum-recorder room on the second floor of the Maryland Street apartment and started thinking through the problem.

He began by listing the three possibilities for the unrest: tectonic, magmatic, or hydrothermal. Under magmatic, he now wrote: *100%*. The COSPEC had clinched it for him. From the magmatic column, he drew two branches: magma would make it to the surface and erupt, or not. From his study on big caldera systems that he had helped prepare for Mammoth Lakes, he estimated there was a 40 percent chance the activity would lead to an eruption. From that estimate, the question was, would it be a big eruption or a small one? Based on the geological reconnaissance work he had been doing, Newhall thought Pinatubo was likely to be big if it did erupt. He estimated the chance of a big eruption to be 70 percent. There would certainly be pyroclastic flows with a big eruption, and Newhall estimated there was a fifty-fifty chance they would flow east, toward Clark and large population centers including Angeles City. While the probability tree had the appearance of quantifiable science, much of it was guesswork. Still, Newhall estimated there was a 14 percent likelihood that Pinatubo would erupt within a year. If it did, there was a 20 percent chance that pyroclastic flows would reach the base and Angeles City. Now, Newhall thought he had what the VDAP/PHIVOLCS team needed to begin raising a warning.

"If we learned one lesson from Armero, and we learned many," he said, "the number one lesson is that it's not enough to get the science and predictions right. You've got to convince people of it. It's remarkable how resistant people can be. There's a lot of denial. We were not panicked [at that time], but we were very concerned about whether we were going to be able to overcome this skepticism soon enough."

Volcanologists walk a fine line in such situations. They have to overcome deep-seated skepticism that the local mountain could be a killer. So the geologists have to generate enough concern to motivate realistic planning for an eruption. But they can't get people so afraid that they will bolt long before an eruption. If that happens, people lose confidence in the credibility of the scientists issuing the warnings. And if the scientists prime a population, warn of an impending eruption, and nothing happens, then people are unlikely to respond as quickly with the next warning.

The officers at Clark were subject to their own pressures. The larger Air Force community in the Pacific questioned whether Clark commanders were being panicked by a handful of ragged research scientists from the States. From Clark, Pinatubo was little more than a bump on the horizon. How could that mound of dirt at that distance threaten the base? The base's other second-in-command, Colonel Dick Anderegg, began getting calls from officers at the Pacific headquarters in Hawaii. More than a few reminded him of what had happened to Chicken Little when the sky didn't fall. "Are you that worried?" they asked.[8]

Although Grime continued to be a holdout, Anderegg and the others were starting to worry. They had been taught by the geologists that ash can be sucked into engines and cause loss of power. They learned it can get into telephone systems, computers, and electrical transformers and fry them. The biggest danger is rain, because ash holds lots of water, which can be dangerous in a country where there are lots of flat roofs.

In one demonstration, one of the geologists took a glassful of ash and dumped in another glassful of water. The ash completely absorbed the water, making the mix as heavy as both glasses. Now, they said, imagine a million glasses of this stuff on a roof.[9] And they played the Krafft video at every opportunity. After these classes, Grime's boss, Major General William A. Studer, finally ordered that a similar education program begin for the entire base.

Power and Lockhart began calling themselves the "base busters." The absurdity of the situation was captured in a line they would repeat over late-night beers: "Hello, General. We're a couple GS-11s, and we're here to close down your base."

It was a joke, a way of relieving some of the tension everyone was feeling. It didn't look like help was coming from the United States. Instead, there were grumblings that this adventure was becoming too expensive to handle. The VDAP mission was to aid PHIVOLCS, but it was becoming increasingly clear that as far as the Air Force was concerned, PHIVOLCS scientists might just as well be cutting grass. And the Air Force was only marginally happier with the U.S. scientists.

In the tradition of the FPP experiment, Chris Newhall decided to have a party. On May 18, he held a St. Helens barbecue. The hamburgers and hot dogs came with healthy helpings of volcanology. Newhall's main concern, he said, was from pyroclastic flows.

"What's a pyroplastic flow?" asked Anderegg.

Not pyro*plastic*. Pyro*clastic*. In eruptions that are big enough, Newhall said, enormous clouds of ash roll down the side of the volcano at speeds of 50 to 100 mph, and since they are typically nine hundred to a thousand degrees centigrade, they wipe out everything in their path. Generally, deep valleys act as flow barriers. But the geologic record around Pinatubo showed the volcano ejected such enormous pyroclastic flows that they jumped from valley to valley, even as far away as Clark.

So what you're telling me, said Anderegg, is that in a big eruption it's possible the people living in base housing could be toast?

Yes.

Anderegg would later write in his journal: "I had no idea how to deal with this information on either a personal or a professional level. What they had just said scared me, really scared me. Not so much for myself or any immediate danger. But how would we manage this thing? How would we make decisions?"

By May 20, the education campaign was paying off and even Grime acknowledged that evacuation plans might be needed. He put Anderegg in charge of a crisis action team, a CAT in military lingo. His directions were to assume the worst. No lives must be lost. All property must be protected. And nothing was to be said to the general public without the base public-relations chief saying it. As for aircraft, they were being flown out as a part of a planned drawdown that had been under way for a year.

When the CAT held its first evacuation-planning meeting, they asked how far they needed to move people to be safe. A geologist spread a large map of the area on the team's conference table. It was 8.2 miles from Pinatubo to The Hill, the residential area at the western tip of the base. He drew a twelve-mile arc that ran through the flight line, and then a fifteen-mile arc that nearly reached the eastern wall of the base.

According to the geological record, which was still being dug up, the entire base was built on pyroclastic flows. But if one looked only at Pinatubo's recent eruptions, there was a 99 percent chance of safety for people at twelve miles, the flight line.

For fighter pilots, 99 percent wasn't good enough because that meant there was a 1 percent chance of everyone being 100 percent dead. No. Evacuees were not going to be statistically safe. They were going to be perfectly safe.[10] It might mean falling back to the U.S. naval base at Subic Bay, twenty-five miles from the volcano and a two-hour drive from Clark. And that might mean moving everyone off base. Clark Air Base might have to be abandoned again.

While the U.S. military appeared to be VDAP's priority, Newhall's main objective was to help the Filipino geologists. They had responsibility for the safety of the over one million people who lived on land that had periodically been devastated by Pinatubo's eruptions.

Newhall and PHIVOLCS were most concerned about the twenty thousand indigenous Aeta who lived on the volcano itself. They were nomadic family groups that ranged freely over Pinatubo, tending their animals, gathering wild fruits, and praying to their god, Apo Mallari, who lived in the mountain.

When Newhall and the PHIVOLCS geologists drove into the field to contact the Aetas, they found that the Aetas, too, were concerned about the earthquakes and the huge, loud steam jets that sometimes threw rocks. They told Newhall that the activity was a reaction to the drilling of the lowlanders. Anyone who didn't live on the mountain was a lowlander to the Aetas, and in the previous two years teams of lowlanders had drilled geothermal test wells around Pinatubo. The Aetas thought this had angered Apo Mallari.

The VDAP/PHIVOLCS team turned to local nuns who operated a mission in the Aeta country and worked with them. These were the nuns who had initially reported the unrest to PHIVOLCS, and they needed little convincing about the dangers of Pinatubo. When the nuns had made their initial report to PHIVOLCS, they had picked up brochures about earthquakes and eruptions, which they had been studying as their mission shook. The geologists and nuns decided not to teach the Aetas about volcanology, which were really just the stories of their own science culture, to convince the Aetas that the volcanologists knew ways of predicting when Apo Mallari's home would become uninhabitable. The geologists promised that the nuns would be warned when it was time to leave and stressed that then the nuns would have to leave quickly.

While Newhall was poking around the camps of the Aetas, he encountered the guerrillas of the New People's Army, a communist group responsible for the recent assassinations of Americans. He and some geologists from PHIVOLCS were working by car on the west side of the volcano when they came to a dead end in a country road that opened onto the village of Belbel. Newhall rolled into the village in a white Chevy Suburban, heavily armored with tinted windows and bulletproof glass, with U.S. embassy plates. Newhall and the PHIVOLCS geologists told the villagers that Pinatubo could erupt, and should they get a warning to evacuate, they should leave the mountain quickly.

As Newhall walked back to the Suburban, he noticed five men who were "straight hairs" or lowlanders. They were dressed as farmers, in shorts and T-shirts, but they were all carrying automatic weapons. Newhall had been expecting to run into them. Coincidentally, the guerrillas had attended the University of the Philippines in Manila, as had some of the PHIVOLCS colleagues traveling with Newhall.

"They hit it off right away," Newhall recalled. "I think they accepted we were not an army patrol, but given the embassy vehicle, they probably wondered if we were CIA. I guess we were sufficiently disarming because they came out to talk to us. And that was very helpful because we could explain to them what we thought was going on, and they in turn could explain it to the Aetas. They started asking us questions about the relationship between the earthquake a year before and this. They asked what is magma and how do volcanoes work. Our message to them was, 'Stay tuned, but if the order comes to evacuate, don't ignore it. Get the hell out.' "

Newhall, Punongbayan, and the head of the Philippine Civil Defense also began visiting the provinces that were threatened by Pinatubo. None of the provincial Civil Defense offices had contingency plans for an eruption. In fact, many were surprised to learn that Pinatubo was a volcano. At each stop, the team pulled out a hazard map to show local officials what kinds of threats they could expect if Pinatubo did erupt.

The mayor of Angeles, which had a population of over three hundred thousand, continued to deny his city faced any threat at all. At one lunch meeting of national and regional officials, the mayor not only failed to appear, he sent his secretary to say he was too busy to attend. The head of Civil Defense was furious and told the secretary that when the mayor was ready to behave responsibly, the mayor should call his office.

By late May, newspaper stories about the unrest at Pinatubo, and the possibility of an evacuation of Clark Air Base, began to be carried in the Philippine press. The reaction in Manila was a lot like that of the mayor of Angeles. Many Filipinos thought the Americans were overreacting.

Others viewed the new talk of a possible evacuation as a pressure tactic to give the Americans leverage for the ongoing base negotiations. Many prominent Filipinos wanted Clark back and converted into an international air hub for that corner of the Pacific. These critics speculated that if the Americans evacuated, it would simply be a demonstration of the havoc that could be wrought on the economy if the Americans were not allowed to renew their lease.

Newhall did not have an easier time with the U.S. embassy. Newhall and General Studer helicoptered to the embassy in Manila and met with the ambassador, Nicholas Platt. The purpose of the trip was to get the ambassador to declare a disaster. The declaration would free up extra funds that Newhall could use to mollify anxious Survey officials back in the States. It would also allow the Air Force greater flexibility in their preparations. Newhall explained that the monitoring data indicated the unrest was driven by magma, and that the geologic record portrayed a history of especially large and violent eruptions. By the end of the meeting, the ambassador agreed to declare a disaster.

It never happened. After the ambassador met with Newhall and Studer, he informed members of the "country team," a group of ten senior U.S. officials in Manila. During the meeting, US AID's frustration with the Pinatubo news stories and the Survey team boiled over. Officials of US AID in the Philippines objected to the Survey's request to declare a disaster yet to happen. To this group, the Survey was a *research* organization. A senior official told the ambassador to ignore the warning because "Pinatubo isn't going to do a thing; it's just a [Survey] thesis project."[11]

"It certainly looked like they were doing a lot of research," recalled Bryant George, who was at the time a US AID official. "The Survey is a research institution. Somebody's got to do it. But it was difficult for us, who see things happen that are easy to predict like a typhoon or a tsunami, to just do what people reading their little equipment . . . are telling us [because] something may happen. We were giving them a hard time because it seemed all they were interested in was one more beautiful piece of research, which didn't mean anything to us because we're the guys who house people and feed people after a disaster. We don't do research. They spent their time bringing in equipment, bringing in bright young guys who looked like they were working on their Ph.D., who wanted an additional chapter or two.

"Remember, the queen here is Clark Air Force Base, which the Filipinos did not believe we would ever give up. And [the Pinatubo scare] was seen as one more stratagem to keep it. We were dealing with a host government who were not convinced that this was anything else than some claptrap, one more thing the Americans had cooked up."

One problem Newhall was able to overcome was the skepticism by the Air Force. Almost from their arrival, the VDAP team had been training a handful of Air Force officers in volcano monitoring. Some spent hours in the Maryland Street apartment. As they became educated, they began seeing for themselves the signs that the volcano was ramping up. Earthquakes continued at a high rate, although they were still confined to the northwest cloud. Perhaps the single most compelling piece of data came from changes in the COSPEC readings. On May 13, Newhall got his first measurement of SO_2 output, which was five hundred tons a day. Ten days later, that had jumped tenfold, to five thousand tons a day.

In the slowing improving atmosphere, Newhall pressed to begin a broad education campaign aimed at the entire base population, rather than just the few senior officers. Newhall was supported in this by Colonel Murphy, the maverick who had started evacuation planing without orders. Specifically, Newhall wanted to appear on the base's television station and play the Kraffts' hazard tape. But Newhall's request was rejected. The senior officers wanted to maintain the chain of command, the formalized relationship between the ranks. The concern was that Newhall or any of the other geologists would make an end run for a decision the officers opposed by going directly to the community.

Newhall did show the video to the senior staff, which, now open to the possibility that the neighborhood volcano could actually erupt, began giving Pinatubo new respect. Newhall thought he could drive home the dangers of a volcano even more if he had the Kraffts come to Clark and show some of their other films of eruptions. He also knew that they had harrowing stories to tell, such as the time they were in a rubber boat only to discover the water was turning to sulfuric acid. Newhall located the Kraffts in Japan, where they were working with Harry Glicken on a project on the Unzen volcano. They agreed to come within the next few weeks, after they finishing filming at Unzen, which was then releasing some spectacular dome-collapsing pryoclastic flows.

While Studer was convinced that Pinatubo could be a serious threat to Clark, he was aware that his bosses at regional headquarters, CINC PACOM (commander in chief, Pacific Command) in Hawaii, had their doubts. At Studer's urging, Newhall decided to leave the Philippines and brief the command of the Pacific military forces. Newhall decided that he also had to convert doubters inside the Survey itself, who continued to pummel him with early-morning calls about finances. So he planned that after he did his briefing in Hawaii, he would return to Reston, Virginia, for a few weeks and lay out his case there. He would leave as soon as VDAP replacements arrived and were able to get a feel for the volcano's activity.

The Air Force and the Navy were now actively preparing evacuation plans. The project was given the code name Operation Fiery Vigil. But the heightened sense of urgency that had overtaken the military was not shared by VDAP. The reason was Long Valley.

"Long Valley was very much a part of our thinking all the way through the time I was at Clark," said Power. "This activity was going on, but it could go on for a long, long time."

In fact, Pinatubo could rumble along for a year or two before anything happened, or before it settled down without an eruption. But the military was not about to be left without U.S. geologists on base to provide expert advice. So, the Pinatubo Volcano Observatory was probably going to be staffed by Survey scientists for months, if not years.

Newhall realized that he would have to begin rotating VDAP volunteer scientists in and out of PVO for as long as it took. Newhall requested a number of replacements, but most of them had long-standing plans for summer fieldwork, and they declined to baby-sit a Third World volcano that might not pop at all. Back at CVO, the new scientist-in-charge, Ed Wolfe, was asked to find replacements for the VDAP crew in Pinatubo. He, too, found it was not easy.

"People are busy," says Wolfe. "They've got their own projects and families. It's hard to ask them to go to someplace for an indefinite period. It's a disruption. At the same time, people do want to do this, but they don't want their time wasted. If they knew that something really big was going to happen, they'd go in an instant. But when it's uncertain and they've got other things to do, it's hard."

Wolfe, who had never attended an explosive eruption, finally decided that if he couldn't get enough replacements, he'd go himself. He was also running up against the meatball/conehead problem.

"The reward system for scientists in the Survey has been one primarily based on publications and the impact of those publications," recalls Wolfe. "If you wanted to get promoted, what you did is do research that enables you to write journal articles. And anything that interferes with that interferes with the speed of your promotions. And going off to deal with a volcano crisis can take a lot of time and may not produce very rich research results. When it comes right down to it, it's the number and quality of your publications that are what gets you to the higher grades. I don't think the people were particularly reluctant to go [to Pinatubo] and participate, but there was an undercurrent of 'I don't get rewarded for this.' "

Newhall again asked Dave Harlow to come to the Philippines. But

Harlow had promised himself that he would concentrate on publications. Newhall's calls to Harlow in Menlo Park were distracting.

Newhall was not the only person lobbying for Harlow's release. Dick Janda, who had recently taken over as VDAP chief, pushed for Harlow as well. On May 15, Janda faxed a request to Harlow's division chief explaining why Harlow was needed. On the subject line, Janda wrote: "Once more into the breach rode Dave Harlow, and . . . ?" On the copy faxed to Harlow, Janda added, ". . . able to walk on pfs [pyroclastic flows], leap caldera wall, survive Bogotá, and calm the troubled waters."

Janda's fax put it bluntly. An experienced seismologist was vital not only to the population threatened by the volcano, but to the safety of the science teams working in the field.

"I can not address the technical merits of Harlow's publication record," Janda wrote, "but I can assure you that during crisis situations interdisciplinary, international crisis response teams feel very comfortable with his judgment."

Harlow fought the urge to go. If he went, he could be put months behind on his publishing schedule. Then he discussed the situation with a friend, Elliot Endo, the seismologist who had first recognized the magma signature in Steve Malone's lab in 1980. Endo convinced Harlow that working on restless volcanoes was his calling. So Harlow phoned Newhall and said he was on his way. As he began packing, he knew this would not help his career in the Survey one bit.

Another geologist, one who understood how to read the stratigraphy of a volcano, volunteered: Rick Hoblitt. Newhall also got a young tiltmeter specialist and VDAP veteran named John Ewert to join the Pinatubo project. Tom Miller, the CVO electronics expert, pledged to come soon, too.

With replacements on the way, taking $40,000 worth of equipment with them, Survey management was beginning to panic. Even if money could be found to replace the equipment, now valued at $140,000, it would take the government procurement process a year to replace it. Unrest under volcanoes in Mexico, Guatemala, Costa Rica, and Colombia had already triggered preliminary inquiries for VDAP assistance. If those inquiries turned into formal requests, the Survey would have to acknowledge VDAP couldn't respond.

On May 21, Newhall got a fax from the VDAP project chief. He wrote that to continue the operation without funding would be "irresponsible."

He concluded: ". . . recommend that if acceptable assurance[s] of funding are not received by early June, the response team should be instructed to start an orderly withdrawal of our staff and equipment."

The Survey's Volcano Hazard Program didn't have adequate funding to support a prolonged, elaborate response without impacting other parts of its program. As at St. Helens, no slush fund was available for such emergencies. The volcano program got a yearly appropriation and it was all spoken for. It was all good work, all important, and all done by dedicated people who had fit their yearly budgets into a multiyear research effort. The branch chief at the time, Pete Lipman, didn't feel he had the authority to go a million dollars in the hole or to clip research projects across the board to fund one international effort at a volcano that might burp along for a year or more and do nothing. There was no Bob Tilling in senior management this time assuring the field scientists the money would come somehow.

Dave Harlow and John Ewert arrived in Manila on May 25. An embassy car picked them up and drove them to Clark. On the way, Harlow stared out the window at the rice fields, water buffalo, and farmers in shorts, rubber sandals, and straw cone hats. It occurred to him that this was the first time he had been in Southeast Asia since his Vietnam War tour.

Because of a deadline for work in Alaska, Power became the first VDAP team member to rotate out. Even that wasn't easy. Flights out of the Philippines were difficult to book. The earliest seat Power could book required a two-week lead time, a 7 A.M. flight, Monday, May 27.

The day before Power was to leave, around dinnertime, the instruments in the drum-recorder room recorded two unusual quakes. The seismometers picked up a long period event. The signal was shallow, in what some geophysicists call the magic zone, close enough to the surface to be dangerous but in the mystery layer so that depth can't be accurately determined. The signature on the drums was unmistakable. It generally meant magma was moving. Then the seismometers recorded the first earthquake underneath the fumaroles, and suddenly, Pinatubo was rocked by a large phreatic explosion.

Base operations was quickly notified of the new activity, and soon Harlow and Ewert were astonished to see a flock of Air Force colonels rush into the Maryland Street apartment, each carrying a "brick," a handheld radio, which they were talking on using code names.

The team at Pinatubo in front of the Maryland Street apartment at Clark Air Base. Left to right: Dave Harlow, Chris Newhall, Ray Punongbayan, John Ewert, Andy Lockhart, John Power, Gemme Ambubuyog (front), and Sergio Marcial (back). Note Pinatubo Volcano Observatory sign, SHAKE AND BAKE WITH THE BEST.

Did the two unusual seismic events mean that the volcano was moving into a new phase, perhaps signaling an eruption was much closer than a year away? Or were the events the first of a series of changes the volcano might undergo as it rattled on month after month?

Power had a simpler question.

"We had some very pointed discussions," recalled Power. "Are things really changing? Nobody was willing to hang their hat on a single earthquake. I remember standing there [in the drum room] and looking at things and thinking, 'Should I stay or go?' I decided to leave and woke up the driver at two A.M. to take me to the airport. It was all based on the idea that we were going to be monitoring the volcano for some long period of time. I remember driving to the airport thinking, 'I'm going to be back the first of August.'"

Power would return before then.

CHAPTER 11

They'll Think
You're a Hero

A fter dropping John Power at the airport in Manila, the embassy driver went to the Hyatt to pick up Rick Hoblitt, who had arrived a few hours earlier. For Hoblitt, the flight had not been pleasant. He had a killer toothache. Worse, the entire family was upset that he was, as he put it, "flying off to another unhappy volcano." Hoblitt had been involved with eruptions at St. Helens, in Indonesia, and in Alaska. But somehow this trip was different, and Rick and his family all felt it. He had a hard time leaving, and his family had a hard time letting him go.

Hoblitt and the other new arrivals were not warmly greeted by US AID either. Ever since the Air Force had twisted arms to get Newhall and his team into the Philippines, US AID complained they had been nothing but problems. For months, the embassy had been under orders to reduce the large official American presence in the Philippines. As the AID staff dwindled, their workload increased. Now with the VDAP team's arrival, and more coming seemingly every day, AID was repeatedly pressed to explain why the U.S. presence was growing.

"We were removing people [from the embassy], some of whose jobs many of us thought were quite necessary," recalled Bryant George, who was the US AID disaster officer at the time. "So the political side had told the government of the Philippines that's what we were doing. Then in the middle of this, the Survey brings in this endless stream of scien-

244

tists and others and they make demands on our logistical capabilities, which did not create the most pleasant of circumstances. One of the guys arrived without a passport. Can you imagine? I mean, he had never had one. He didn't leave it at home. God, I could have killed him."

By the last days of May, the new arrivals were all in place or on their way. Lockhart had decided to stay to work on seismology with Harlow, although Lockhart's recurring sinus infection flared up. He kept working in the field, nursing electronics and trying to keep repeater stations operating. But the infection was grinding him down. One of the first things Hoblitt did after he arrived was to address Andy Lockhart's biggest concern: Was Maryland Street built on a pyroclastic flow deposit?

The quickest way to determine if pyroclastic flows had swept across the base was to examine the riverbanks along the north and south sides of the base. Soon, Hoblitt and a couple of young PHIVOLCS geologists were wandering through shantytowns, splashing across streambeds and sewers, scraping the walls of river cuts, and using a magnetometer on the deposits he found. He was doing what he had started doing years ago at St. Helens. He was examining the magnetic orientation of the deposits to determine if they arrived as pyroclastic flows or hot mud. Mudflows could wreak as much havoc on the towns along the stream as pyroclastic flows, but dangerous mudflows would take an hour to reach the area around the base. If the deposits were pyroclastic flows, it meant that eruptions could reach from Pinatubo to the base in a few minutes. The only way to survive those was to be watching from a distance. There wouldn't be time to run.

Along one riverbank that ran just adjacent to the south end of the runway, Hoblitt found a deposit ten to fifteen feet thick, colored a mix of tan and gray. It was what geologists call poorly sorted, which means, a wide variety of grain sizes were in the deposit, like plum pudding. Being poorly sorted meant that whatever put this deposit in place, either a lahar or a pyroclastic flow, was highly energetic when it reached this point. It illustrated that an enormous amount of energy had driven past eruptions. Hoblitt was convinced this particular deposit was a pyroclastic flow—at a stunning distance from the volcano. When Hoblitt and the PHIVOLCS team finished their work, after dark, they concluded both hot mudflows and pyroclastic flows had swept through now populated areas that bordered Clark. By inference, they also believed Maryland Street was built on pyroclastic flows.

The Air Force was informed that the entire base, not just the residential Hill housing area closest to the volcano, might be swept by deadly flows in an eruption. The chance was small, but it was not zero.

Still, by the end of May, the VDAP team had no indication that Pinatubo would have another eruption anytime soon.

Seismologist Dave Harlow was not even convinced the activity was caused by moving magma. To him, Pinatubo's unrest was primarily geothermal. In his mind, the big earthquake of a year earlier had shaken the piping system of the volcano. The result was to be seen in the huge, billowing white clouds jetting from the fumaroles on the summit. The earthquakes, even the long period event on May 26, could be explained as a geothermal system acting up.

As for the COSPEC readings, Harlow didn't have much faith in them. While high gas readings had signaled a lot of eruptions, many times SO_2 levels didn't correlate. St. Helens, for example, never had any SO_2 above background levels before May 18. And at Galeras in Colombia, huge amounts of SO_2 were released without a major eruption.[1]

To help clarify the situation, John Ewert began deploying his tilt stations.

While traveling outside the base to install tiltmeters and to replace dying car batteries at the seismic stations, Hoblitt, Ewert, Lockhart, and four PHIVOLCS scientists drove through Crow Valley, the site for the mock air wars. It was a world of targets—airports, jets, bunkers—built of wood and wicker. Marines were stationed on Bunker Hill to guard the microwave relay stations, which also relayed the signals from several of the seismic stations. Guards were needed because, in this impoverished corner of the world, people regularly scavenged for material to sell. The Air Force had little doubt that if they abandoned Bunker Hill, the scavengers would cart off everything from generators to the blocks in the walls of the guard tower. In fact, the driver of the Suburban stopped when a Filipino approached the vehicle with a ten-pound bombshell over his shoulder. Hoblitt, who had some experience with artillery in the military, saw it was a live shell and yelled to the driver, "Get the fuck out of here!"

With the military preparing its evacuation plan, Hoblitt suggested it was time for the team to begin looking at its own plans for an eruption. To monitor the eruption while it was happening, a rare scientific opportunity, they would need a fallback position. Just as before, they had a difficult time finding a safe spot with a line of sight to the instruments and a view of the volcano. To be sure the fallback site wouldn't be overwhelmed by a pyroclastic flow would mean relocating off the base entirely. That would probably be unacceptable to the Air Force, and with power supplies uncertain at any time off the base, it was probably not practical.

So the base would have to do. They looked for a spot as far away from Pinatubo as you could get that was still on the base, which meant they needed a site somewhere around the airstrip on the east leg of Clark's triangle. Hoblitt thought the air-traffic control tower would have been fine, since it had a good view of the volcano, but the Air Force was not about to turn over such a vital station to the geologists.

They found a compromise spot beyond the tower and even beyond the runways. It backed right up against the north-south wall on the east side of the base. Known as the Dau Complex, it had been a communications center during the Vietnam War. The single-story, cinder-block building was an almost perfect observatory. Dau had its own generator, and an air-filtering system, windows that faced the volcano, and a working air-conditioning system, which was stuck permanently at beer-cooler temperature. They immediately began wiring the place. In case they needed to abandon Maryland Street in a hurry, they could quickly activate the communications center without losing too much data.

Preparing a fallback position had an additional benefit. Those preparations did more than anything else to convince senior officers that it was time to take Pinatubo seriously, said Anderegg. The Air Force began instructing its personnel about what to do in the event of an evacuation.

Around the beginning of June, the volcano was sending confusing signals. The seismicity was changing. While the northwest cloud remained just as active as ever, a second cluster of earthquakes began appearing under the steam vents. The quakes were hard to recognize and difficult to locate because they were so small. But Harlow, with the skill of a master, identified them and pinpointed their source. To him, the tiny quakes were moving along a line that, taken together, looked like a conduit opening beneath the fumaroles.

The fumaroles themselves were changing. They were jetting much higher than they had been, up to a thousand feet above the summit sometimes. And they were getting gray as the plumes began to include rock as well as steam. The ash was tested, but it was made only of old rock fragments. There was no sign of fresh magma.

Then, the amount of SO_2 drifting out of the volcano plummeted. From a high of 5,000 tons a day on May 28, it dropped to 1,800 on May 30, then 1,300 tons on June 3, and 260 tons on June 5. The volcano was clearly changing character, and the VDAP/PHIVOLCS team was not happy about it.

"As long as the volcano was freely degassing," said Hoblitt, "that's a good sign. When it stops, then you have to start looking for an exit. It could mean that it has turned off, but it could also mean, especially if you have a high SO_2 flux that goes very low, it probably means it's pressurizing."

Pressurizing, to a volcanologist, means the mountain is doing the same thing as a pressure cooker. It's building up pressure inside a mountain of rock.

When the SO_2 emission rate drops precipitously, it often signals that the channel the gases had been flowing through is sealing up. The seal forms because magma becomes more viscous when gas separates from it. The top of the magma column, where gas can most easily escape, becomes a seal, preventing or retarding the escape of gas from below. The pressure often increases until the strength of rock is exceeded—then a gush of gas-rich, low-viscosity magma will break through the seal and stream upward. So magma migrates upward in a series of starts and stops, and this causes activity to wax and wane.

How quickly, Hoblitt wondered, could a volcano move to a huge eruption?

The answer to that had implications for the warning system that was being used by the Survey and PHIVOLCS and was now being adopted by the Air Force. It was a five-step ALERT system, developed on-site:

LEVEL ONE—low level of unrest, no eruption imminent.

LEVEL TWO—positive evidence of magma involved; could eventually lead to eruption.

LEVEL THREE—high and increasing unrest; eruption possible in two weeks.

LEVEL FOUR—intense unrest with long period quakes; eruption possible in twenty-four hours.

LEVEL FIVE—eruption under way.[2]

Both PHIVOLCS and the Survey were at level two. To Hoblitt, the system was based on the belief that the scientists would be able to provide warnings of two weeks and twenty-four hours before an eruption.

What is the possibility, Hoblitt began to wonder, of an explosive eruption without additional warning? As he knew from his first experience with another volcano, it was possible to be overconfident. Yes, a lot had been learned in the last decade. Computers gave the scientists a power they could not have dreamed about at St. Helens, but experience, too—from Indonesia to Alaska—told them that each volcano was unique. On June 2, Hoblitt faxed a memo to Newhall, who was in Hawaii preparing to brief the Pacific command.

"It has occurred to me," Hoblitt wrote, "that the current alert-level system implicitly assumes that there is a 100 percent probability that we will be able to provide adequate warning of an impending eruption. While I agree that the chances of this are good, I also think that there is a nontrivial probability that an eruption will occur without any warning, or with only a little warning. . . . My best seat-of-the-pants estimate is something between 1 percent and 10 percent. If you find this paranoid, blame it on St. Helens."

The next day, Newhall called to discuss Hoblitt's memo. Harlow was on the extension in the drum room, listening in, when his monitors begin flickering. A unique magma-associated earthquake, a long period event, rolled in, and then another. Reports about activity began being shouted throughout the apartment. A big quake shook all the stations. It looked on the instruments like an explosion. Ewert, looking out the second-floor window, saw lightning over Pinatubo. Maybe Pinatubo was erupting. Harlow checked the monitors again and saw harmonic tremor, another likely signature of magma. It, too, was being recorded on all stations. Somewhere in all of this, Newhall had been disconnected.

Within minutes, Air Force officers were rushing into the Maryland Street apartment. The geologists recapped the events for the military and said they were thinking about going from level two to level three. Harlow and Hoblitt recommended against it, but they warned that the volcano was revving up.

For breakfast the next morning, Andy Lockhart whipped up a batch of pancakes. When the phone rang, Andy took the call. It was a colleague at CVO. There had been an eruption at Unzen, he was told. Maurice and Katia Krafft and Harry Glicken were missing.

Within hours, it was confirmed that all three had been killed. The Kraffts, Glicken, and forty Japanese journalists had gone to Unzen for its dome-building eruptions. High on a hill across a valley from Unzen, they set up a viewpoint that allowed them to see the dome. The valley between the observers and the mountain had diverted all previous pyroclastic flows away from the site. But the day before the call, a large part of the dome had collapsed and pumped out an unusually big pyroclastic flow. A portion of the overriding cloud, lighter than the main flow, broke off and surged straight up the valley wall and over the party. Although this flow was extremely dilute and left ash just a few inches thick, it was strong enough to carry a taxi away and hot enough to kill everyone almost instantly.[3]

Hoblitt called Dan Miller, who was back at CVO. Hoblitt said to his old friend, the person who had charmed him into dropping out of chemistry, that he was concerned. "Here I am playing with another loaded volcano, and I've just lost three friends," he said. Dan asked if he needed more help sent out there and Hoblitt replied, "I don't think there's time."

In Hawaii, Newhall and a colleague from HVO briefed the officers of the Pacific command. Newhall, his beard long and clothes a bit tattered, was briefing the starched Navy and Air Force commanders. He gave them the same briefing he had been giving on Clark. Pinatubo had a history of big eruptions. Since April, it had been steaming and quaking like an awakening volcano. In mid-May, SO_2 readings began climbing to high levels before ominously dropping off the scale. Recently, the quakes had been registering signals that indicated magma could be moving. All of this was consistent with a volcano heading toward eruption.

Then he explained his probability estimates. There was a fifty-fifty chance of an eruption within a year, and a 10 percent chance that an eruption could impact Clark Air Base. And, just so there was no confusion, eruption didn't mean slowly rolling magma, such as they had here in Hawaii. It meant blast zones that strip jungles, pyroclastic flows with searing hurricane winds, and fast-moving mudflows that could sweep away buildings. "These things are real," he said. To drive home his point, he played the tape the Kraffts had made, which included scenes from the pyroclastic flows that had destroyed Saint-Pierre, from the directed blast of St. Helens, and from the mudflows of Armero. When it ended, he told the officers that the people who had shot most of this film knew as much about the dangers of eruptions as any volcanologist in the world. And yesterday, they were surprised by a volcano and it killed them.

The next day, June 5, Lockhart, in growing misery with a sinus infection, helicoptered with one of the PHIVOLCS geologists to Negron to replace a failing battery. The day was clear, and on the return, the helicopter swung by the fumaroles. The vents were in the bottom of a narrow, deep valley. On another flight two days earlier, Hoblitt had noticed an unusual feature at the base of the fumaroles, but then the wind had changed and visibility was obscured. Today, Lockhart looked at the base of the steam vents and saw a spine.

A spine is an extrusion of lava that is so rigid that it can literally stand up like an obelisk. It means the volcano has reached another milestone on the way to an eruption. Technically, with lava actually coming out at the surface, it means an eruption is under way.

The science meeting that night was filled with bad news. Lockhart detailed his observation of the spine. They discussed a paper published by Newhall, which reported that about half of all explosive eruptions are preceded by a spine. The SO_2 levels recorded that day were at 260 tons, practically nonexistent. The instruments at UBO and PIE were showing a rapid acceleration in the number of quakes.

Adding up the observations, Ray Punongbayan decided to raise the warning to level three, eruption possible in two weeks.[4] About ten thousand Aeta tribesman were to be moved from their homes to evacuation camps. Most of the U.S. geologists agreed with the decision to go to level three.

When two Air Force officers dropped by to see if anything was happening, they were given the news. Harlow called the base commander, Colonel Grime, and told him there could be an eruption in the next two weeks and it might put stuff on the base. Harlow agreed to brief General Studer and his staff the next morning at 6:15 A.M. The scientists wanted another look at the spine. They called ops to request a helicopter, which required some greasing, but they got the flight. Before Hoblitt went to bed, about 10:30 P.M., the drums were scratching out a series of long period quakes.

Hoblitt woke at four-thirty and was on the flight line by 5 A.M. With preflight checks, the chopper didn't lift off until 5:40 A.M. But the weather was clear and in less than ten minutes Hoblitt was looking out the side door for the spine at the base of the fumaroles a few hundred yards away. He spotted it and looked again. The chopper made two more passes before Hoblitt was certain. Andy was wrong. The spine he saw was really an erosional remnant, the wall remaining between two enlarging vents. It was old rock, not new lava. Hoblitt remembered that Harlow was to brief the general about raising the alert level based on the appearance of a spine. Hoblitt told the pilot to rush back to the base. The second they touched down, Hoblitt jumped out, grabbed the nearest phone, and managed to get Harlow just as he was walking in to brief thirty officers and the region's commander. Most of them had already heard about the spine and the higher alert level. Hoblitt told Harlow that there was no spine. "Have a nice day," Hoblitt said, and hung up. After the meeting, Harlow said, "I don't think I enhanced my standing very much."

In the briefing, Harlow had to back off the spine observation, but he didn't back off to level two. The Survey and PHIVOLCS were at level three. The system had been devised with fixed time periods attached to each level. Levels could be raised at any time, but it would take two weeks before they would ratchet down a level. It was done this way because volcanoes rarely ramp up in a straight line, but in a pattern similar to patterns in the stock market. It's the trend that's important, not the day-to-day blips. To prevent geologists from being lulled by temporary quiet, it had been decided to take two weeks before a warning level could be lowered. So now Harlow, Hoblitt, and the rest of the geologists were stuck with level three. Punongbayan was not unhappy about it because it allowed him to enlarge the evacuation areas.

The episode cost the geologists some of their hard-earned credibility. In fact, the Air Force decided they would devise their own alert scheme. They were not about to leave decisions about the safety of personnel or the evacuation of the base to people who couldn't tell a pile of rock from an eruption of lava. But there was another reason they decided to have their own alert system. They were not prepared for an evacuation. No one had been informed where he or she was to go and what to bring, and more important, Subic was not ready for fifteen thousand guests.[5] The U.S Air Force volcano alert level was set at two.

For Dave Harlow, the day had been a disaster. No one talked about it much, but Harlow's Vietnam days seemed to be catching up with him. He was a grunt Marine in the sixties, and like many returning veterans, he carried a lot of troublesome memories home with him. He had been angry with the military for two decades. Now, the grunt was briefing the general and his staff, trying to walk a line that all volcanologists find difficult, between appropriate alarm and thoughtful restraint. More and more, he drifted off to his bedroom and pulled a pillow over his head. John Ewert began to kid him about it. But Harlow was feeling the heat, especially from the base commander, Colonel Grime.

"He [Grime] was just pissed off that this whole thing was happening," says Ewert. "And by extension, he was pissed off at us. He didn't want to deal with a volcano. He didn't want a bunch of rogue scientists on his base. He just didn't want to bother."

Dawn on June 7 was a little brighter for Harlow when he checked the overnight seismic readings. The number of quakes under the volcano was up overall. UBO, the instrument nearest the vents, was really hopping. But more important, the number of quakes beneath the vent was roughly equal to those in the northwest cloud. There was also the beginning of what could become a convergence of shallow and deeper quakes. John Ewert was watching the readings from his tiltmeters flow into the drum room, and he was now convinced that the volcano was erupting. But Harlow was a seismologist and felt most comfortable with seismic data. Even observations can let you down, but the quakes can't lie. Harlow believed that he was beginning to get a feel for this volcano, and that it was heading toward eruption.

Hoblitt and Ewert took the dawn patrol and reported back that the vents seemed to be more vigorous and more ash seemed to be lying around the vents. When he returned, Hoblitt called one of Grime's staff, Colonel Murphy, whose job was essentially to manage all the base's infrastructure. Hoblitt told him the seismicity was up. Murphy wanted to know what the hazards boundaries were, and Hoblitt said the chances of a lethal flow reaching the base were small but he couldn't guarantee absolute safety on the base. Before Murphy hung up, Harlow said that he wanted the backup facility wired as soon as Murphy could get the phones in, and that had better be soon.

Grime had been summoned for a noon briefing at Maryland Street. Harlow ran down the day's data. The number of quakes was up, and during the morning, the location of those quakes had been rising.

Activity at the vent was up and seemingly more ash was being ejected. Ewert's tiltmeters told him the volcano was inflating. Hoblitt ran down his newest hazards maps. He said there was a 1 percent probability of pyroclastic flows reaching the flight line, the farthest point on the base from the volcano, about twenty kilometers away from the vents. The scientists also told Grime that they were preparing to fall back to the Dau Complex. Grime, who couldn't ignore the volcano anymore, said only, "We're fucked."

After St. Helens, many of the geologists would say the one thing they needed most but didn't have was time to think. In the frenzy of a crisis, there are too many phone calls, too many meetings, too much fieldwork and educating and faxing and arranging for flights, and repairing shaky equipment, and arguing for more money, and too little time for eating, sleeping, and thinking. Now, in the middle of this crisis, Hoblitt was able to put the demands of the moment aside and let questions bubble up to occupy his thoughts.

The first thing he wondered was why there were two clusters of quakes. It was uncommon. Volcanic quakes can move from place to place. Often, they begin at a distance from the volcano and migrate to the spout. But two clusters of quakes occurring at the same time was unusual. It was a feature new in Hoblitt's experience and, as far as he knew, new in the literature. Yet there, on the computers, were the signals—three miles apart.

Hoblitt began to wonder if the quakes were related. Could they be part of one body of magma? If the quakes were on the ends of one underground plug of molten rock, Hoblitt calculated that would mean there was a magma lake beneath Pinatubo of about fifty cubic kilometers. The rule of thumb in the volcano business is that a volcano erupts about 10 percent of its magma. Here, that translated into an eruption that would produce five cubic kilometers of materials. And that was a big number. It was ten times larger than St. Helens. Five cubic kilometers was bigger than anything Hoblitt or even Harlow had seen. It could be bigger than any living geologist had seen. It was Krakatau size. But it was also consistent with Pinatubo's past eruptions. Hoblitt explained all this at one of the science meetings attended by Colonel Anderegg and other senior officers.

"I remember sitting there and listening to this and getting a chill," said Anderegg. "We all looked at each other and said, 'We may not be anywhere near far enough away from this volcano.'"

At 2:20 P.M., General Studer arrived and the Survey team ran through the same briefing.

"If you were God," Studer said looking at Hoblitt, "what would you do under the present circumstances?"

This is the question the Survey volcanologists had always tried to avoid answering. They believe that calling an evacuation is the job of the person in charge, either the one elected or appointed. The scientists advise. They can describe what the volcano has done in the past, what it is likely to do in the future, and the probabilities of when it might do what. But the call is not part of the assignment. It was the same at St. Helens, and it has been with every crisis VDAP has worked. The geologists had decided the call was a question of acceptable risks, and only local officials can know what risk is acceptable in their communities. More important, the local decision maker can be held accountable for his actions long after the Survey scientists have moved on to another volcano.

But the rule against providing advice, while clear and simple, is fuzzy in practice. The local decision makers work side by side with the geologists over long hours and under stress. It builds a bond, and when a friend asks, "What should I do?" it's hard not to answer the question. And it's just seductive and flattering to have a two-star general ask, "What should I do?"

Hoblitt reluctantly said that he would start moving dependents off base. Studer left the briefing without indicating his intentions.

Hoblitt felt he was on a roller coaster. Events were progressing rapidly. In the afternoon, Harlow, upstairs in the drum room, shouted, "These events are getting shallower." At five o'clock, the volcano ejected a steam plume that reached twenty-eight thousand feet, the highest so far.

Harlow said he was ready to call a four, eruption within twenty-four hours. He phoned Punongbayan. Newhall had already heard the news and phoned Punongbayan urging him to pull back his geologists from their observation posts northwest of the volcano. Punongbayan left his people in place but raised the alert level to four, which enlarged the evacuation zone.[6]

Grime and Studer were informed that the Survey was moving toward a level four. Grime returned to Maryland Street at 5:14 P.M. Harlow said that because of the continued seismicity, he was about to call a four. Grime wanted to see the drums himself. Grime said, "You know that once you declare a four, the end of the world happens." He meant that the Air Force had set level four as a trigger for Clark to be evacuated.

By the time the officers' cars pulled away, the seismicity had plunged. Harlow's heart began to sink. Harlow said, "We are at a high three." After Grime left, Colonel Murphy arrived and quickly sensed Harlow was backing off. He told Harlow to stand firm or Studer would use his equivocation as an excuse not to evacuate. But as Harlow stood looking at his monitors, he saw the entire seismic network calming. Harlow called Punongbayan and told him the seismicity had backed off and maybe a four was premature. Punongbayan was mad. Another Air Force colonel arrived. He said he had been sent by Grime to "take care of any needs" the VDAP team had, but he had obviously been sent to monitor the geologists. Ewert noted the tilt had dropped back to its pre-inflation level. Hoblitt concluded they had just had a small eruption and scribbled in his notes, "Jesus—we're riding the tiger."

Harlow was beginning to think that his credibility with Grime and Studer was completely shot. After the previous day's briefing about the nonexistent spine, Harlow had run up to a level four and then both the seismicity and the tilt had dropped off a cliff. For Harlow, it was turning out to be another bad day.

"We had this very large swarm [of quakes], which is clearly a magmatic swarm," Harlow recalls of that day. "It had a whole different flavor to it than a main-shock/aftershock sequence. This was a swarm that started out slowly, built very strong, and I placed a call to Colonel Grime and General Studer. By the time they came over, the thing was over. It was over. The thing was back at background level. It was another example of something happening and we got everybody all excited, and then it quieted down. About this time, some of our relations with the military were getting a little strained."

Providing guidance to local decision makers about how the neighborhood volcano is behaving is often a roller-coaster ride. Generally, volcanoes don't move from dormancy to catastrophe in a straight line. More often than not, they rev up and drop back, rev up and drop back, although they rarely drop back to background levels. But most people expect things move in a straight line. This jerky dynamic can be a hard pattern to get adjusted to, not only for the local mayor or colonel but for the volcanologists as well. It's especially difficult because any single episode of ramping up can progress to a full, catastrophic eruption. And any single repose can lead to full dormancy. Geologists often personalize this cycle by saying the volcano is toying with them. But it exacerbates the general tension, adds to the personal stress and sense of danger, and makes sleep even more difficult.

"There is a tremendous amount of concern about calling things right," says Ed Wolfe. "At Pinatubo, there was a lot of concern that the volcano could toy with us, go on and on, threatening but not doing anything. We were concerned about overstating the problem to the Air Force and losing our credibility. That was the nub of the issue. It's knowing this was a very dangerous volcano, that there's extreme unrest going on, and yet not really being able to say, 'Yes, it's going to erupt within the next week.' The sense was that the Air Force was willing to take mitigating steps, but by God something better happen. So it put a lot of pressure on the team."

A complicating factor was that the geologists were not just concerned about getting an "eruption" call right. They felt they had to get a "no eruption" call right as well. The challenge was understanding and communicating the *appropriate* level of concern.

"The culture we have developed," said Wolfe, "is that we're just as concerned about causing a lot of social disruption unnecessarily as we are about having lives lost. We're trying to be right on the mark. We don't want to have false alarms. We don't want to have big evacuations that turn out to be unnecessary. And at the same time, we don't want to lose lives either. So we pay a lot of attention to our credibility, to try to have a pretty good basis for the pronouncements we make, and to try to have the people who have to act on those pronouncements understand what the evidence is. And that's true in all volcano responses."

Ewert, who had been oddly happy in the middle of this bleak situation, told Harlow he'd done the right thing to call a four. Don't worry, he said. Harlow headed back to bed with a pillow over his head.

Harlow's mood was directly linked to the action on the seismographs. John Ewert, likewise, was linked to his machines, the tiltmeters. Based on what his tilts had told him, Ewert was convinced Pinatubo had just put lava on the surface. He outlined his evidence at the evening science meeting. Most of the others found it hard to believe. They were certain the day's rainstorm had jiggled Ewert's instruments or the concrete was warping or a million other things.

Tiltmeters are so notoriously untrustworthy that frustrated geologists created the Golden Rules of Tiltmeters. Rule number one, don't use tiltmeters. Two, if you do use them, don't believe them. Three, if you believe them, don't publish. And four, if you publish, don't be the first author on the paper. Andy Lockhart says tilts can be valuable, but there's a "huge bogosity overhead."

There are several problems with tiltmeters. First, they are too sensitive. They measure tilts in microradians. A microradian is the tilt that would be made in a rigid bar, six-tenths of a mile long, when a dime is inserted under one end. Tiltmeters take a long time to settle in after they have been planted. They are set in cement, and it can take weeks for the cement to cure. During that period they are unreliable. Also, electronic noise can affect the readings. The warmth of the sun or rain influences the tilt measurements. Even Ewert is sensitive about the noise in tilt readings.

"It's a rule of thumb for almost anything you do with a volcano," he says. "You never want to trust one parameter with your life and prognostications. You're always looking for a number of parameters all pointing in the same direction."

Ewert believed he was doing just that when he said his instruments were indicating lava had come out on the surface. His data fit the seismic data perfectly. In his view, the volcano had inflated. The inflation caused the increase in both tilt and seismicity. Then, lava broke the surface, perhaps with the high steam plume. Both the tilt and the seismicity died off precisely because the lava extrusion had relieved the pressure on the system. He had seen the same thing at St. Helens. Pinatubo had put lava on the surface, Ewert was convinced of it.

Later that night, Ewert had to go to Manila to pick up Tom Murray, coming in from CVO. Before he left, he stuffed his day pack with two sets of clothes and packed his shower kit. He thought the mountain might blow at any minute and he wouldn't be able to make it back to Maryland Street.

At 4:30 A.M. the next morning, the phone rang in the Maryland Street apartment. Pacific command in Hawaii was erupting. "There's been a fuckup," Hoblitt wrote in his notes that day. "Studer and [Air Force] at

[Clark Air Base] didn't know we declared a four. They found out from above in the chain of command. Apparently when a two-star general gets word from a four-star general, it makes for an unhappy camper." Hoblitt woke Harlow and asked, "Didn't you tell Grime we were at a four?" Harlow, groggy, said maybe he didn't communicate effectively. The phone rang again. Lockhart grabbed it. It was Pacific command again. They wanted to know if the Survey was at a four or not. Hoblitt told Andy to have them call back in five minutes.

"Okay," says Hoblitt, "we *are* at a four. The decision was sound when we made it. We must accept the blame for the fuckup. Air Force careers hang on it and our relationship with the Air Force."

Lockhart passed the word along to Hawaii, then turned to Harlow and said, "Welcome to the penalty box."

At 6:30 A.M., the dawn patrol took off to look for Ewert's lava but the summit was enveloped in clouds. Ash emissions were the heaviest they had seen. The team flew to the BUGZ seismic site and replaced a battery. While they were working, three young Filipino men wandered by, and Hoblitt told them they were in a danger zone and should leave. They didn't seem too concerned. A little before 8 A.M., the chopper left BUGZ and made another pass north of the summit. The fumaroles sat at the bottom of a steep river canyon on the volcano's north flank. The canyon and fumaroles were obscured in the clouds of ash. Briefly, the valley cleared, and Hoblitt saw a dome some distance from the fumaroles. Ewert was right. A big, gray, ugly glob of rock had poked out the east valley wall. Hoblitt asked the pilot to make several passes, just to make sure this was no mirage. It was a dome.

Hoblitt returned to Maryland Street and called Newhall, who was back at Survey headquarters in Reston. The appearance of a dome meant one of two radically different things. One camp believed that when you see a dome, you can relax because the magma has degassed. The energy that puts the explosion in explosive volcanoes has been drained off. If the magma is gas-poor, it means a big, explosive eruption is unlikely. Volcanoes are, after all, gas-driven machines. No gas, no boom. This blob of lava they saw this morning was obviously gas-poor or it would have come out as flying foam rather than oozing rock.

The second possibility was that only the dome lava had degassed as it neared the surface. It could be the tip of a rising magma tongue that could be highly gas-rich, and therefore explosive. Newhall and Hoblitt decided this was probably the correct interpretation, since it fit Pinatubo's history. Also, the seismicity was back up at a high rate, which was just the opposite of what had been seen at St. Helens's dome-building eruptions. Pinatubo was still breaking rock.

Newhall asked Hoblitt if he could get a sample of the new rock. It would mean flying into the canyon where the steam vents were jetting and occasionally blasting ash up five miles. Hoblitt said he didn't think the return would be worth the risk. A sample would tell them the magma composition, including the gas content. This would be a guide to the explosive potential of the volcano, and this could be valuable data. But going into the steam valley would just be too dangerous now. They had Pinatubo's history to tell them about the mountain's explosive potential; they would just have to wait to see exactly how explosive it would become.

In the old days, when eruptions were once-in-a-lifetime events, Hoblitt or any other volcanologist in this situation might have risked dashing into the valley. But the days of the volcano cowboys were drawing to a close. In one sense, these scientists had seen too many eruptions and lost too many friends. To be sure, explosive volcanoes still held their fascination. But now they knew the power of this beast and gave it a respectful distance. In part, they did it because the odds were beginning to work against them. If there was to be only one eruption in a person's lifetime, then the odds were on his side. But the more eruptions one worked, the shorter the odds became. So when Hoblitt declined to fly into the valley, he signaled a new era had begun. The volcano cowboys now had families and mortgages and they knew there would be more loaded volcanoes in the future.

Still, for Hoblitt and the other VDAP members, risk came with the job. Nosing around an active volcanic vent was not safe. The deaths at Unzen were an inescapable lesson.

There were still risks to take. Some of them would be carefully weighed, and some decisions would be taken on too little sleep and information. At times in the future the pull of the data would override their best judgments. And they would stay in a hazard zone if it meant they could gather better information to save lives. Ruiz was another lesson. Maybe they had become volcano smoke jumpers, diving into an unknown risk to do a dangerous job because, in part, it was a social good and, in part, because they loved the big show.

Technically, with lava on the surface, Pinatubo was in eruption. It was not much of an eruption, but the geologists were happy as clams. Hoblitt wrote in his field notes, "The dome saved our butts."

About noon, General Studer and Colonel Grime arrived at PVO for the day's bad news. Harlow said the earthquakes in the northwest cloud had disappeared and the quakes under the vent had grown shallow. The interpretation was that lava had pushed close to the surface. It was not even breaking rocks in the northwest cloud. But the seismicity had fallen off, so Harlow told the officers that the Survey was at a "low three." The general said the Air Force was now using the same scale as the Survey and they, too, were at a three, which meant they had started moving pregnant women and other vulnerable personnel off base.

General Studer said he was concerned about the morning's communications problem. "We fucked up," said Hoblitt, which raised a smile on General Studer's face. Hoblitt promised that the next time Studer would be informed decisively and without equivocation. Studer looked at Hoblitt and said that they would evacuate the base when Hoblitt's gut told him it was time to go. Hoblitt replied his guts "aren't calibrated well enough to know if a given period of activity will progress to a catastrophic event." Studer said he had confidence in Hoblitt's gut.

As the officers began leaving the observatory, Studer said that he had the ultimate authority for the situation and to call him directly anytime and not worry about disturbing him. Grime joked that the scientists haven't worried about bothering them yet.

Almost as soon as the officers departed, a convoy of Philippine military vehicles pulled up outside the Maryland Street house. Out stepped several Philippine generals, who were immediately encircled by the mob of Philippine journalists they had brought with them. With cameras rolling, Harlow and the crew were questioned about the situation.

Unlike at St. Helens, the Survey geologists were operating outside the reach of the media. They had been insulated from the press by the guarded wall surrounding Clark Air Base. This was the way they preferred to operate. It was how Chris Newhall wanted it. To him it was preferable because in a foreign country they wanted the local geologists to be the spokespeople. They were the ones who could best relate to the community, and they were the ones who would be left behind to explain what had gone right or wrong.

It was also preferable to most of the geologists because they viewed the press as a distraction. The job of the scientist was to educate the decision maker. Journalists, in their view, were often poorly informed and vulnerable to manipulation by anyone with *geologist* printed on his or her business card. Reporters could not separate the significant from the inconsequential, they thought, and the result too often was a poorly served public that was either overly anxious or dangerously complacent.

After the Philippine officers and media departed, Harlow and Hoblitt went next door, to a second town house they had acquired, to eat lunch in peace. But just as they sat down, Lieutenant Colonel Al Bryant arrived with an aide. Bryant ran the bombing range at Camp O'Donnell, which was the location of millions of dollars of electronic equipment. Bryant wanted to know how safe his troops were. Hoblitt estimated that since O'Donnell was nineteen miles from the summit, his troops were relatively safe, but they could be hit by mudflows. How much of a warning can you give us? Bryant asked. Hoblitt explained that for mudflows, he could give relatively long warnings, but they would also include more false alarms. Or he could give a more confident warning closer to an eruption. That was the pickle.

"This volcano shit is just about the limit," Bryant said. "My troops have been assassinated, the base has been shut down by labor strikes, there is a revolution going on in the hills all around us, and now this volcano is threatening to erupt."

"You're fortunate to have so many rich life experiences," said Hoblitt, and both men laughed.

That night, the seismographs were reporting that the northwest cloud had become active again. Hoblitt couldn't sleep. The phones were ringing, people were talking in the drum room, and his mind was whirling. He got up and did his laundry. Checking the drums, he saw that seismic activity was picking up.

At 6 A.M. that morning, June 9, Hoblitt was at the flight line when Ewert called to say there had been an explosion, followed by a long period event and volcanic tremor, all signs of moving magma. About ten minutes after takeoff, the helicopter was over the summit. Heavy ash emissions were rising from the vicinity of the dome, but the dome itself was not visible through the cloud. There were no gross changes in the volcano's edifice.

Flying over the northwest and north flanks, Hoblitt saw no evidence of pyroclastic flows, just a hell of a lot of ash, much of it carried by the wind to the northwest. It was hanging in low areas, in the vicinity of the northwest cluster. As the wind blew it around, it looked as if it were coming out of the ground. If that was the case, it could mean that magma had reached the surface at the other side of a large magma body.

This led Hoblitt to consider the possibility that he could be witnessing the early stages of a caldera-forming eruption. If he was right about the size of the magma body, the eruption could be ten to one hundred times the size of St. Helens, and that's a lot of material pumped out of the ground. Hoblitt was now envisioning a process that was typical of such gigantic eruptions. After a volcano has ejected a large amount of material in a vertical eruption, the space that once was the magma chamber becomes partially empty. The roof rock above the deflated chamber is no longer supported by the underlying magma, and it falls into the emptying chamber. When that happens, hot gas and ash rise along fractures in a circle miles away from the central vent. If a caldera formed at Pinatubo, it could produce pyroclastic flows that would easily blast through Angeles City and Clark.

But Hoblitt soon realized that the ash had been carried from the volcano's summit and fallen to the ground, where it was being blown about by surface winds. Still, the possibility that Pinatubo could produce a caldera-forming eruption didn't disappear on those winds.

The helicopter headed back to Clark, but instead of landing near the airstrip, the pilot sat the craft down in the middle of the base, on the parade field near the headquarters building. Air Force security people rushed out to the chopper and fitted a blue bucket seat into the helicopter, which was soon filled by General Studer. Colonel Grime was at his side. Studer wanted to see for himself what was worrying the geologists. Not more than five minutes later, the helicopter was back hovering between the northwest cloud and the summit, near what Hoblitt believed to be the largest plug of magma to approach the surface of the earth in this century.

"God, that's a lot of ash," the general said as he eyed the plume rising from the valley.

"That's nothing," said Hoblitt. "Look down there."

From that platform, it was easy to understand Pinatubo's threat. The topography of the volcano and the land around it were like a Japanese parasol, with Pinatubo's little summit sticking up in the middle. The earth sloped gently from Pinatubo for tens of miles in all directions.

Hoblitt pointed out that the slope of the land was formed by millions of tons of pyroclastic flows and volcanic ash. If you squinted and blurred your vision, Hoblitt said, you could more easily see that everything sweeping out from the center of this parasol is volcanic debris, hundreds of yards thick. Over the centuries, rains had cut through these deposits, forming narrow, steep canyons, some of which were a thousand feet deep. The height of those walls was an indication of the size of Pinatubo's eruptions. They were monstrous. And that was the only size eruption Pinatubo had produced.

"Now, General," Hoblitt said, "look down that valley."

From this height, the valley Hoblitt was pointing to looked like the sight on a gun barrel, a notch cut through the volcanic sheet. It sloped due east from the top of Pinatubo. It was the path pyroclastic flows from Pinatubo had always taken to reach what was now Clark.

"See that ramp there?" asked Hoblitt's voice in the general's flight helmet. "See where your base is on the edge of that ramp?"

Studer continued to gaze out the helicopter's alcove. He nodded and stared down the ramp. Eventually, he turned to Grime and said, "Do it tomorrow. Do it tomorrow. They'll think you're a hero. All the kids will get out of school."

When he returned to PVO, Hoblitt was told that PHIVOLCS had gone to a five.[7] PHIVOLCS observers on the west side of the volcano thought they spotted pyroclastic flows. The radius of the evacuation zone was extended to twelve miles, and the number of evacuees increased to about twenty-five thousand.[8]

The warning level system was now a mess. The Philippine government was at five, the Survey at four, and the Air Force still officially at three. The problem, of course, was that while the Survey scientists were issuing the best possible guidance, the two groups of decision makers were reinterpreting those results for their own ends. If the Air Force went to a four, it was evacuation time, the "end of the world," and there was great reluctance to abandon the base. But the main concern for PHIVOLCS was giving people enough time to move away from the volcano, and Punongbayan wanted as much time to do that as he could get. To the VDAP team, PHIVOLCS was overplaying the eruption possibility while the Air Force wasn't as concerned as it should be. Both were dangerous. If the volcano didn't blow shortly after PHIVOLCS called a five, then in time people would start moving back into the danger zone, and when the next warning was sounded, some of them would ignore it. The danger for the Air Force was that personnel would be on base if Pinatubo let loose an unexpected eruption and, since the base was built on volcanic flow deposits, that could be catastrophic.

That night, the VDAP team splurged on takeout from Pizza Hut. Newhall called from the States and was relieved to hear the base would be evacuated.

At 6 A.M. on Monday, June 10, 1991, the order was broadcast over the military television and radio stations to evacuate. The residents had been prepared. For days, the base newspaper and television station were broadcasting details of the evacuation plan—what to bring and where to go. The personnel were told to pack enough for a three-day trip. Their homes, they were promised, would be kept secure by the 960 base policemen who would be remaining. Also remaining would be a few hundred cooks, engineers, communications specialists, and heavy-equipment operators.

Immediately, the streets of Clark began to fill. Neighborhood by neighborhood, precisely as planned, cars, trucks, and buses funneled downhill and formed lines on the runway. From there, they were escorted off the base, through the surrounding shantytowns, and onto the federal highway for the hour-long trip to the U.S. Navy installation at Subic Bay. By noon, only six hours after the first announcement was made, over 14,500 people had left Clark. Just 1,200 remained.

Studer called the evacuation in the face of strong skepticism from Washington. In newspaper stories carried in the *New York Times* and the *Washington Post*, the mayor of Angeles City had said the Americans were "overreacting" and "causing a panic."

When geologists travel to a new volcano, it is a habit for them to pack their trunks not only with tools of the trade but with all the relevant research papers they can quickly lay their hands on. Now, Hoblitt reached into that floating library and began reading about column collapse.

A column collapse, which produces something called a pyroclastic surge, was first appreciated because of nuclear bomb testing in the South Pacific. In 1946, military planners wanted to see what a nuclear explosion would do to a naval fleet. So, in July of that year, they detonated a nuclear bomb underwater in the lagoon of the Bikini Atoll. The explosion produced the familiar nuclear column, composed largely of water droplets and gases. This aerosol column collapsed and began moving out from the base at hurricane speeds, sinking ships. In 1962, another nuclear test, called the Sedan test, was triggered 650 feet beneath the Nevada desert test site and produced a similar column fall-back and surge. The two tests showed that fast-moving surges were possible over water as well as over land.[9]

A famous example of pyroclastic surges occurred in the Philippines in 1965, documented by HVO alum Jim Moore. Early in the morning hours of September 28, people living near the Taal volcano, thirty-five miles south of Manila, were awakened by earthquakes. They found Taal fountaining brilliantly with an incandescent lava spray. But Taal developed a crack near its vent, which allowed water to mix with the magma. When that happened, the eruption began looking like the Bikini Atoll test, except this time the base surge was filled with searing gases and rock fragments. The surge blasted at hurricane speeds, snapping trees and sandblasting trunks. Almost 200 people died in the eruption.[10]

When an eruption is big enough, about the size of the one the VDAP team was expecting from Pinatubo, the eruption column can be spectacular. Material is ejected from the vent at enormous speed largely by the action of pressure release, like taking your thumb off a shaken bottle of Coke. But the eruption column is also pulled upward by the thermal effects (hot air rises) of the ash cloud.

Very fine particles can remain in the atmosphere for months. But what concerned Hoblitt this night were those bigger particles that fall back down in a mass as a blistering cloud of dense hot gas and ash, like what happened at Lamington. At the base, the debris falling back can spread out rapidly along the ground as a pyroclastic flow or surge. It was late when Hoblitt finished reading, took two Sominex, and fell asleep.

When Hoblitt and Lockhart climbed into the helicopter for the following day's dawn patrol, the pilot told them the service people had been getting calls from their families at Subic. It was not pleasant news. Subic had already been eight hundred beds short before anyone arrived from Clark. The extra 14,500 people were straining all the resources of the base. Hundreds of evacuees were billeted in unoccupied, unfurnished apartments, and others were living in officers' and enlisted-personnel clubs. People were sleeping in the gymnasium, classrooms, chapels, and hallways. Hundreds more had been adopted by Subic families. Every space with a roof had evacuees under it. Still, thousands remained without shelter.

Many waited for hours in field bleachers, under the Pacific sun. Some men dug holes in the sandy soil under their cars and crawled into the shade. Thousands were left to sleep in their cars, which thieves attempted to burglarize at night. The pets quickly became a problem. They couldn't be left in the cars and they couldn't be allowed to run wild over the base. The Marines wanted to euthanatize the animals.

The urgency of the situation was growing by the hour. Within days, the typhoon season would begin battering the Philippines with lashing winds and rain. If this evacuation was uncalled for, the helicopter pilot told the VDAP team, there would be sore feelings to deal with when the families returned. The chopper's blades begin whining, the helicopter lifted off, and it hit the team what it meant to have their advice followed.

Although the geologists were confident that Pinatubo was heading toward eruption, the increase in activity that had resulted in this evacuation might not lead to an immediate eruption. In their worst scenarios, the final catastrophic eruption might not come for weeks. In between, Pinatubo could easily wind down its activity and then ramp up, again and again.

If it was tough on the evacuees, it was tough on the geologists. More than a year later, one of the PVO geologists would write in an after-action report one of the most candid assessments of this observational science: "This was a stressful time for the small team of PVO scientists. We were concerned about the volcano and the safety of those around it, including ourselves. We were also concerned about the serious consequences of a false alarm and whether we would have a second chance should the volcano not erupt as anticipated. Sleep was difficult, nerves were taut, and we were at our physical and emotional limits."

The geologists had wanted the base evacuated. They felt Pinatubo was fast approaching a crisis, and they wanted people inside and outside the base at a safe distance. Now, nearly everyone was at a safe distance. This would not be another Armero.

But it was all based on Ewert's unreliable tiltmeters, Harlow's mystical impressions of the seismicity, Newhall's yo-yo sulfur dioxide readings, and Hoblitt's hasty interpretation of the quickening pace of unrest. Because of all that uncertain science, the largest military base the

Ironic sign advertising the current movie. (PHOTO: RICK HOBLITT)

United States had overseas had been evacuated. Beyond the base, hundreds of thousands of Filipinos had their lives disrupted and their livelihoods put at risk.

"We had all of those times when nobody paid any attention to us," recalled Harlow. "And finally somebody does, and you wonder, 'Is this Long Valley, Mammoth Lakes? Is this going to be another one of those times when we really embarrass ourselves?' I couldn't help second-guessing myself. All of us did. I was feeling as though the chances were pretty high that we would all be hauled in front of committees investigating the disastrous evacuation, its cost and impact to the Philippine economy and on the Air Force in general. Suddenly those visions were dancing in my head."

The base was nearly empty. Streetlights came on after dark, but no families walked around the parade grounds. The Pizza Hut and Baskin-Robbins were closed. No one was at the bar in the officers' club. But outside the club, a sign advertised in bright red letters that this coming Friday, June 14, the film would be *The Last Days of Pompeii*. A theme party would follow.

CHAPTER 12

Eruption

What had once looked like a small town in Middle America suddenly seemed to be a town under martial law. There were no children on bikes. All the parking slots were empty. Every store was closed. There was no sound of kids or lawn mowers or music or television. Now, every few hundred yards a security cop in full battle dress stood with an automatic weapon.

With the base evacuated, the VDAP crew decided it was time to put some distance between themselves and the volcano. They began transferring monitoring equipment to the backup facility, the Dau Complex. Dau was as far down the hill as they could get and still be on Clark Air Base. It was even beyond the runways. It snuggled up to the long eastern wall of the base, but it still seemed too close to the volcano.

"Dau wasn't really where we should have been," recalled Ed Wolfe. "Under different or better circumstances, we should have been farther away from the volcano. But there wasn't a good alternative available to us, and we had great support from the Air Force, which was really important. So we were closer to the volcano, within a potential hazard zone, than we wished. I don't think anybody was confident a pyroclastic flow could not reach us."

Dau was only three miles farther away from Pinatubo than the Maryland Street apartment, but it added a margin of safety. The geologists had calculated that a pyroclastic flow traveling at 225 miles per hour would take about three minutes to reach PVO. They estimated it would take another minute or so to sweep through Dau.

The geologists transferred their rows of small computers and recorder drums from Maryland Street. Because the VDAP team chose to sleep at the Maryland Street apartment rather than shiver sleepless in Dau, they left one drum at the apartment to monitor things during the night. They began an around-the-clock watch at Dau.

The geologists shared Dau with a contingent from the military. Rooms were designated as sleeping quarters for the general, the other officers, and the Air Force operations personnel, and a small space was designated for the geologists. Another sleeping area in an airplane hangar was to be used almost entirely by the Air Force security police who had remained behind to safeguard the base from looters.

While everyone else seemed to be fraying from lack of sleep, Dave Harlow was oddly alert. A survival habit he had picked up in Vietnam had returned. When life there was intense, infantrymen often worked shifts of two hours on, two hours off. After a while, grunts learned to nap at every opportunity. With naps, Harlow found he could go three or four weeks without a full night's rest and still be alert, even on night patrols. When the other geologists had seen Harlow with a pillow over his head, they'd assumed he was tormented from the briefings and foul-ups. Actually, he was napping. But he didn't tell his fellow geologists about the naps. "I was embarrassed to talk about it because it seemed a little bit weenie-ish," he said years later.

With the base evacuated, the geologists were able to resurrect one seismic station, the first one planted on Clark, designated CAB. After it had first been turned on, the geologists had found that it was too close to a water tower. It seemed that every time someone flushed a toilet, the instrument recorded it. But with the base evacuated, Andy Lockhart realized that water wasn't being drawn down or pumped up. Lockhart was sinking further into his sinus infection, but in an effort to be productive, he replaced the battery in CAB and got the station on-line again.

Harlow was especially grateful to have CAB back on-line. When the base was evacuated, the Air Force pulled its men out of Bunker Hill. Shortly after they left, a truck bounced over the rough roads into the high-tech valley, and those inside it stole everything at the base, including the generator. With the power gone, Harlow lost the relay station that bounced signals from three of his instruments. With the remaining seismometers, he could plot the number of quakes and their intensity, but he was unable to figure out their location or depth. The loss frustrated Harlow unlike anything else he had suffered at Pinatubo, from unpleasant Vietnam memories to botched briefings. He was upset because it meant the team would be blind, in one sense, and because a more complete network could have provided an enormous amount of data for study later. Geologists don't get to monitor significant eruptions very often, and when they do, they want to squeeze out every drop of data and make every observation they can.

Frustrations and tensions began to build after the June 10 evacuation, largely because the mountain behaved tamely. But on June 12, at 3:30 A.M., Andy Lockhart woke up, stumbled into the bathroom, and then decided to check the remaining seismic drum at Maryland Street. Holy shit! Activity is way up. Andy wakes everyone and calls Dau to see if the night crew has seen it. The duty person says he just picked it up. Everyone bolts downstairs, jumps in the Suburban, and tears downhill for Dau. Looking in the rearview mirror, they can see lightning coming from Pinatubo. Is that an eruption? At Dau, radar operators call to say they've got a thunderhead on their screens that doesn't look quite normal. The cloud top is up to thirty-eight thousand feet. The geologists begin talking about what they will do if things get too hot in Dau. The Air Force has its own fallback site at an agricultural college farther east. They also talk about the possibility of ash being carried to this side, the east side of the volcano. The prevailing winds are now toward the west, but at this time of year prevailing directions change from west to east.

By sunrise, Pinatubo is belching ash to as high as fifteen thousand feet, and the radar returns are stronger than ever. Seismicity had dropped just after the three-thirty fire drill, but is now building again.

Hoblitt and Lockhart get to the flight line by 6:30 A.M. and lift off for Pinatubo. Again, the helicopter drops down onto the parade ground to pick up General Studer. By 7 A.M., the chopper is on its way to the volcano. As soon as they rise, Hoblitt and the others can see that there has been an eruption that has dumped ash over Pinatubo's eastern slope, the side facing Clark.

They fly out to take a look at UBO, the instrument just east of the summit. UBO had stopped transmitting at 10 P.M. on June 11 just as an earthquake occurred. This morning, hovering near the geothermal drill pad upon which UBO is set, the geologists can see that the whole area has been heavily ashed. The ash doesn't appear to have flowed down from a higher vent. It is as if somewhere near UBO a vent has just popped open and blown a cloud of ash over the countryside. As the helicopter moves closer, to within a couple of hundred yards, the chopper's scanner picks up UBO's signals. UBO is still functioning, but its signal isn't reaching Clark. Perhaps the ash has been hot enough to melt the cable that runs from the instrument to the antenna. As they hover, Rick and Andy discuss whether it is safe enough to land and repair the cable. They decide not to land. If they are on the ground when more ash pops, it will certainly kill them.

In the windy helicopter, Hoblitt picks up a familiar scent and realizes Pinatubo has just raised the ante. It's the smell of cooked vegetation, which Hoblitt first encountered at St. Helens. He turns to Studer and says the volcano has ejected its first pyroclastic flow. He points to a spot in the Maraunot River, which channeled the pyroclastic flow. The hot debris has now come to rest with boulders steaming. They fly back to Clark. Hoblitt, Murray, and Lockhart discuss whether it's safe enough to try to get some of the seismometers back on-line. They decide UBO is too dangerous, but they will try to get PPO back on-line.

At 8:50 A.M., Hoblitt is on the phone in the drum room on Maryland Street talking to Ray Punongbayan. Hoblitt is briefing Punongbayan about the pyroclastic flow and the other events of the morning when something catches his eye. Hoblitt looks at the drum that sits in front of the window and then out at Pinatubo. "Gotta go!" he says, and slams down the phone. He yells to Andy downstairs, "Let's go!"

"About that time," Lockhart recalls, "I open the door on the apartment, stepping out with Hoblitt just behind but rapidly accelerating past me, and I see this huge cloud go up from Pinatubo, and I'm thinking, 'Holy shit!' Hoblitt comes flying past me and we race to the truck and get into this thing. I was just shitting nickels at this point. It was a bright clear day and you could see this huge thing. It's bigger than anything I'd ever seen before, and it was moving really fast. It was going up, but I knew that the distance it was going up was probably about the same distance it was expanding across to where we were. At any

The eruption of Mount Pinatubo on June 12, 1991, as seen from Clark Air Base.

rate, it was kind of terrifying and Hoblitt was remarkably cool about it.

"We went over to the briefing area, which was near the landing strip, about where the control tower would have been. Hoblitt says, 'I got to go in here and tell these guys that this thing has happened.' I'm thinking, 'Fuck these guys, let's get to the Dau Complex.' But stopping is probably the Christian thing to do, although I wasn't very much in favor of it. So we stop there and Hoblitt kind of jogs in and comes right back out. It turns out they knew that it had happened. I couldn't conceive of anyone sitting inside and *not* having known this had just occurred. Hoblitt's got his camera and he stops to take some pix. I just about shit. I yelled at him, 'Hoblitt, Jesus Christ, let's get the fuck out of here.' I'm sure I ruined what would have been great pictures for him. We got back to the Dau Complex, and it wasn't until we got there that I started to feel safe. I just felt that extra three miles, and especially three miles where you could see it, you know, see the whole space because we put the width of that airfield between us and Pinatubo, and I just felt a whole hell of lot better. It was clear by that time that it was going up, and nothing bad was coming out the side. It looked like a big-ass vertical eruption, and it was spectacular as all hell."

When Hoblitt and Lockhart screech to a stop in front of the Dau Complex, they see Harlow dancing. It's Dave's eruption jig. He is happy as hell. They have made the right call. Even if nothing else happens, thinks Harlow, we have still done a good job. The visions of a public humiliation before Congress go up with that eruption.

When the Air Force officers got a look out the headquarters windows at what a volcano could do, many found it "terrifying."[1] A jet pilot has a special feeling for how objects move to altitude. An F-15, one of the fastest combat aircraft in the world, will turn a pilot into stone as it rockets to six thousand feet in about a minute. Pinatubo's cloud got to that altitude in five seconds. And it kept going. By the time an F-15 hits forty thousand feet, it is straining against gravity, and even at full throttle, its rate of climb has slowed from six thousand to fifteen hundred feet per minute. Pinatubo hit forty thousand feet in a little over thirty seconds and wasn't slowing. As the pilots stood watching the cloud blast through fifty thousand feet without pausing, one of them said, "If we could climb like that, we'd be unbeatable." At twelve miles, the hot cloud slammed into the cold layer of air between the upper atmosphere and the troposphere and began expanding into a roiling, circular mass. Clark's officers no longer had any doubt about the power of nature. Any doubt vanished beneath the cloud that spread over Clark Air Base. It was shaped exactly like a nuclear bomb's mushroom cloud. And it had been created in absolute silence.

When Harlow finally went inside Dau, he saw that thirty-eight minutes after beginning, the activity on all drums had dropped to their lowest levels since April. It could mean this was the main event. But given the seismic energy that had been building up, Harlow thought more was to come.

Now, just as they were moving into the climax of the eruption, equipment problems grew worse. Not only had they lost a chunk of their seismic network, the COSPEC was lost. The Philippine Air Force had become more helpful as the events progressed to a level five, and Punongbayan had taken advantage of that to arrange to fly the COSPEC on an Air Force training plane. The small plane, a version of the Cessna 172, was sitting on the tarmac at Clark Air Base on the morning of June 12. PHIVOLCS scientists had just lashed the body of the instrument into the baggage compartment and the COSPEC sight was sticking out, pointing skyward. The plane had taxied out and was preparing for take-off when the mountain erupted. If the eruption had happened ten minutes later, the geologists and the pilot on board would have been part of the eruption column. When they saw the cloud climbing, they decided there were safer places to be than at Clark. They continued their takeoff but turned away from the mountain and headed toward the safety of Manila, with the instrument still on board. It would take weeks to get the COSPEC back to Pinatubo.

Cory Aquino, the Philippine president, helicoptered up from Manila to see the damage, to get a briefing from the PHIVOLCS geologists, and to make appearances before the press. General Studer decided to cut back the force on the base. He evacuated six hundred of the remaining security and support personnel, leaving six hundred on the base.

By afternoon, the scientists at Dau had a new worry. They had gone over the morning's records but didn't find any indication that an eruption was coming. Everything in the seismic record and on Ewert's tilt instruments looked perfectly boring. Then zap! This was what Hoblitt had worried about all along. Blame it on St. Helens.

The scientists had made the call that an eruption would develop within two weeks; then, within twenty-four hours. It was a significant advance from the St. Helens days. Because of the warning, hundreds of people were not now lying dead around the volcano, nor were hundreds of thousands in a panic trying to flee from the mountain at this very moment. And a billion dollars of equipment had been moved out of harm's way. Still, the geologists wanted to make that call closer.

But each volcano has its own style, and the geologists were just learning Pinatubo's. One thing was clear. Pinatubo wasn't the type to back off. While the trend lines may have been a little jagged, the climb was steep. Whenever Pinatubo made a turn, it was always for the worse. So whatever Pinatubo had in store next would likely be big and would probably come soon. That recognition made everyone even more edgy and sent Harlow for a nap.

About six that evening, the seismometers close to the volcano began jumping. Ewert's tilt was showing rapid inflation. Hoblitt called the radar center and was told big thunderheads were around Pinatubo, but there was no eruption profile.

"Keep looking carefully," said Hoblitt. "The drums are beating."

Seismic activity began to ramp up. Most important, the long period events increased for two hours, until Pinatubo released another blast at 11 P.M. Radar operators were stunned to see the eruption column leap off their scope at eighty thousand feet in fifteen seconds. Hoblitt and others rushed outside. They couldn't see anything but a sky full of lightning.

Hoblitt woke up at 5 A.M. on June 13 and his first thought was about whether a cryptodome was forming. It would account for the high level of earthquakes and the drop in SO_2. He bounced the idea off another geologist, who didn't buy it, since the open vent blasting ash and fresh rock would be letting off pressure. Hoblitt said that would be reassuring, except that Bezymianny had produced vertical eruptions before it unleashed a directed blast from a cryptodome. Hoblitt decided the best way to check it out was to go take a look. If it looked as if a fist were trying to punch its way out of a pillow, well, that wouldn't be good. As Hoblitt left for the flight line, long period events were scratching out their signature on the drums.[2]

On the flight line Hoblitt talks to the helicopter crew. He explains that he is worried that Pinatubo is forming a dome underground that could lead to a dangerous eruption. Hoblitt says he needs to know how badly the ground around the vent is deforming, and the only way to do that is to go look. If a cryptodome is forming, it could lead to a massive eruption. If that happens while we're out there, he says, we're toast.

Before Hoblitt finishes discussing the risks, Colonel Grime arrives and starts inquiring about the pilots' rest time. The helicopter crew has only had three and a half hours off. Grime says that the second crew will be fully rested and ready to go at 10 A.M. Hoblitt agrees that will be fine, but he writes in his field notes, ". . . hope [waiting] is a good decision."

It is. At 8:41 A.M., Pinatubo releases its third eruption in twenty-four hours. Like the others, this one blasts through eighty thousand feet without pause and flattens out in a cloud that again covers the base. It looks even more like a nuclear bomb than yesterday's blast.

"How does it feel to have a nuclear weapon go off right outside your base?" Ewert asks General Studer.

"I don't like it very much," he says.

Studer wants to know if it is safe to bring dependents back from Subic to collect items from their homes. When they were evacuated, the message was that residents would be back home in three days and that's how they packed. Now, it's beginning to look as if they won't be able to return for two weeks at least. They want to get clean clothes and to retrieve the irreplaceable items of their lives. Hoblitt is reminded of the Spirit Lake cabin owners who wanted to get to their places on May 18. Now the seismic record is consistently showing the signature of long period earthquakes up to two hours before an eruption. If that is Pinatubo's pattern, the geologists have just learned how to provide a two-hour margin of safety. Studer decides to allow the evacuees to return in stages, provided they are able to make it back across the flight line five minutes after the warning siren begins to whine.

Shortly, the first busload arrives. As each person is dropped off in front of his or her residence, a red cone is left as a marker. Just as the last person is dropped, Pinatubo pops. There was no warning. The bus races downhill, stopping at each cone and lurching to the next.

Studer tells Hoblitt that the Pentagon is putting pressure on him to move his operation to the fallback position and abandon the base. He doesn't want to go. Hoblitt says Studer can tell Washington that the geologists don't think it is time for them to abandon the base, and they can't see how they can stay on top of the situation at a greater distance.

Life in the Dau bunker is becoming surreal. When Andy is out in the field, he finds a huge, dead spider. He wraps it up and brings it back to Dau and carefully places it inside a cardboard tube of cheeseballs. He puts the lid back on the container, puts the container on a nearby desk, and waits. Finally, Harlow comes along, opens the cheeseballs, looks inside, and puts the lid back on and sets the container back on the desk. Andy, who had been fishing for a sucker, is amazed. As Harlow walks away, Andy thinks: "Dave's got bigger problems."

They are sharing space with Philippine barbers who are keeping the Air Force officers neatly clipped. Frogs and ants have invaded the hallways. Almost everyone has orange fingers, since the PX was raided after the base was evacuated and someone liberated a year's supply of cheeseballs. Chomping on the cheeseballs seems to relieve some of the anxiety. The meal option is an MRE, a meal ready to eat. These are multicourse, high-calorie meals wrapped in thick brown plastic. Hoblitt finds them quite tasty. They are often packed with little surprises, such as miniature bottles of Tabasco sauce. Hoblitt plucks out a packet of jelly from one MRE, holds it up, and says to the VDAP team, "Put jelly in your pocket, because if we have a worst-case eruption, we're toast."

By dinnertime, an eruption pool has been created. For $1, bettors get their pick of a fifteen-minute window. If Pinatubo has its cataclysmic blast then—jackpot. All the geologists bet the big show will begin within the next few hours. Hoblitt picks 10:30 to 10:45 P.M. The next morning, June 14, Hoblitt is up at five-thirty. Nobody won the pool, so the money has rolled over for a second pool. Ewert, who has been monitoring the activity during the night, tells Hoblitt that Pinatubo is really restless. Harlow's gut is telling him, "Today is the big one." Hoblitt picks the 7:15 A.M. to 7:30 A.M. slot and drives over to heliops (helicopter operations). He intends to stay on the ground until Pinatubo pops. An eruption is imminent and it would be too dangerous to fly before it goes, but when it does, he plans to jump into the air for a better view. His 7:30 A.M. window passes without an eruption.

Everybody is waiting for the climactic eruption. Fuel trucks are moved into hangars. Air Force combat photographers have their cameras loaded. Hoblitt phones Dau and is told the total energy of the quakes is the highest it's been and the average magnitude of the quakes is also at its highest. The pool has rolled over again. Nobody expected it would go this long.

"We were just beside ourselves because this thing should have erupted already and it didn't," recalled Ewert. "It just keeps banging and banging and banging, and you know it's really building up to something big."

The base meteorologist alerts the Dau warriors that a typhoon is bearing down on Clark. Typhoons are Pacific hurricanes, and this one is expected to hit with winds over one hundred miles an hour. The weather forecasters expect typhoon Yunya to hit tonight. The refugees are now caught between an erupting volcano and an oncoming storm.

Hoblitt again grows concerned about a cryptodome forming. The lesson of St. Helens is hard to shake. He isn't concerned about a St. Helens lateral blast. The geometry of the mountain just isn't the same. But St. Helens did tell him that even something as huge as a volcano can pack a surprise.

Hoblitt is particularly concerned about the possibility of a Lamington-type eruption with its column collapse and base surge. The conditions that produce such an eruption are poorly understood, but it may involve a cryptodome—a shallow pool of magma that can accumulate within or just below the surface of a volcano. The magma is under great pressure because of the weight of the overlying rock, because new magma is forcing its way in, and because the gas in this new magma is coming out of solution. The pressure can of course be relieved by eruptions. But the way it's relieved is critical. If magma flows upward through a narrow channel, the pressure would be relieved gradually, and a typical vertical eruption would result. But if a wide channel suddenly opens, the magma pool will be expended in one great gush and the resulting ash fountain will go up some distance and collapse. The collapsing cloud will send out pyroclastic surges in all directions at high speeds.

If a cryptodome has formed, large-scale deformation of the volcano might be visible, but it's been days since he's seen the volcano from close range. Hoblitt tells General Studer he wants a look at the summit. He hasn't seen it in three eruptions and it could have undergone substantial change. It's a risk, he says. But with the typhoon bearing down on them, this may be the last time he can get a look at it. Studer agrees to the flight, but only if he goes. He won't put his men under a risk he isn't willing to take himself.

At 12:08 P.M., they are airborne. They fly up the Sacobia River and skirt the chain of craters on the northeast flank. Hoblitt can see no new cracks in the ground. The helicopter nudges its way closer to the vent until Hoblitt can see the dome. Everything around it is gray. From the dome's center, an insignificant wisp of ash rises into the clouds. The dome has grown into the shape of a large cow pie and spreads across the river valley. But for now the area around the vent shows no sign of gross deformation. It relaxes Hoblitt somewhat. The chopper lands at Clark at 12:38 P.M.

Harlow and Murray are at the landing zone. Harlow is as anxious as he has been since arriving at Clark. He has lost three seismometers to the vandals who trucked away a generator. With the three instruments he has left, he can't pinpoint the location of the quakes, and it's driving him nuts. Here he is, looking at what is building up to be one of the eruptions of the century, and every second the volcano is putting out data that Harlow can't capture.

He wants to fly back and rig a system to beam the radio signals back to Clark. Hoblitt feels the risk of his flight this morning was worth it because if a cryptodome was forming, then perhaps Dau and maybe Angeles City should be evacuated. It was, in a sense, a civil defense effort. But Harlow wants to go and actually land in the hazard zone to reestablish the link between Dau and his monitoring equipment. This isn't only a risk to save lives. It's also a risk to do science. This is why Dave Johnston was willing to perch on the ridge of Coldwater II, even though he was afraid of the volcano. It was why the Kraffts and Harry Glicken had gone to investigate Unzen. Eruptions are the only experiments volcanologists get. Eruptions are those rare opportunities scientists have when the beasts come alive. To be unable to pull from them as much information as possible is a tragedy. It breaks faith with all those who have gone before.

"I was just overcome by the loss of the data," Harlow recalled. "We couldn't get any decent location on those earthquakes, and they were big. It would have given us tremendous insight on how these systems evolve after the first vertical blast and all those individual eruptions leading to the major eruption.

"Finally there was a break in the activity and I said, 'Man, let's get out there and fix this.' There was a fair amount of resistance to doing this because of the obvious dangers, because the thing could really erupt then and threaten anyone out there attempting to work."

I'm going, Harlow says. The winds are picking up and thunderheads are moving in. At 1:09 P.M., Pinatubo launches another eruption, but it's not the big show. When the eruption cloud begins to diminish, Hoblitt says if they are going, the best time to go is just after an eruption—now.

The plan is to redirect the signals to the highest point on Clark Air Base, the control tower. The job is more risky than aerial observation because it means setting the chopper down and mucking around on the ground. It's a vulnerable position. The pace of the eruptions is quickening and Pinatubo has pumped out some big pyroclastic flows already. The repair crew might be on the ground when the big one goes. Or they could be in the air and be overcome by a cloud of ash. Either way, they would be toast.

The repair crew—including Harlow, Murray, Hoblitt, General Studer, the pilots, and the combat photographers—lifts off just before 2 P.M. The crew is in constant radio contact with Dau. The vent is still pumping out copious amounts of ash, although the seismicity has dropped down, as it has right after most of the eruptions. Thunderheads over the volcano are mixing with the ash, so visibility is marginal. As they approach, the winds begin pushing the helicopter around and lightning is flashing between the clouds. Hoblitt spots something he has never seen before, a bolt of clear-air lightning. They land, but the pilots keep the rotors spinning. Murray and Harlow get to work redirecting the transmitting mast. Studer is on the radio to Clark, listening for word of building seismology or other pre-eruption signals. Hoblitt watches the continuing eruption column, which looks as if it's building again. Winds increase, Hoblitt notices, driven either by the volcano's thermals or by the approaching typhoon, or both. Thunder is constantly rolling from the summit and through the canyons.

Andy Lockhart radios that he is not receiving the station's signal. Harlow and Murray again try repositioning the antenna and plug in a new battery. Still no luck. Thunderheads are building around the vent. Hoblitt says the plume seems to be broadening. At 2:35 P.M., they give up and take off.

Within minutes after they leave, Pinatubo releases its first big ground-hugging surge of ash.

The winds have shifted. Ash is now heading toward Clark. The helicopter can't return to Clark and is running out of fuel. They put down at a remote military outpost, find a car, and drive back to Clark. By the time they return it's raining like hell. Inside Dau, the seismograph needles are slamming rail to rail.

Hoblitt gets a report that lahars are filling the rivers just beyond the north wall of the base. Hoblitt and two PHIVOLCS geologists drive over, and long before he arrives, he hears the roar of boulders clanging together in the muddy river. The air also has the St. Helens smell, meaning pyroclastic flows have cooked vegetation upstream.

At ten o'clock, Hoblitt is asleep in one of the Dau dorms, curled up inside a down sleeping bag because the place is still cold as a meat locker. At 11:40 P.M., the lights go on and someone yells, "We are in a major eruption! Pyroclastic flows are moving toward Clark!"

Bleary-eyed, Hoblitt scrambles into ops. There is talk of evacuating the base to the agricultural college. Anderegg says to Harlow that Pinatubo can't keep up this pace for long. Harlow simply points to the computer monitoring the total amount of seismic energy being released. It is five times higher than it was when Studer ordered Clark evacuated.

The control tower reports a pyroclastic flow has stopped short of the base. Hoblitt wants to see for himself so, under a wild lightning display, he drives over to the control tower.

The tower has been monitoring pyroclastic flows with an instrument known as Cyclops. This is an infrared telescope that the Air Force took off a fighter jet. It was placed on top of the control tower to help direct the Air Force police toward thieves coming over Clark's walls in the night. When the geologists learned that there was a way of seeing the volcano at night in infrared, they drafted Cyclops into service to look for pyroclastic flows moving toward Clark. A hot line was set up from the control tower to Dau, and Cyclops was rigged with a videocassette recorder to tape the events.

The tape Hoblitt sees is disorienting. Cyclops rotates as it moves, so the video's image spins. But Hoblitt can clearly make out several elongated strips of hot spots. What could produce those, he wonders.

He takes the tape back to Dau and plays it over and over and over. All the geologists stare at the images, sometimes turning their heads in unison as if they were watching a tennis match in slow motion. Hoblitt watches the tape long after others head for their cots. At 4:45 A.M., he gives up and goes back to bed himself.

While he sleeps, the seismic instruments are recording continuous long period events. Fluid magma and its gases are moving beneath Pinatubo.

Less than hour after going to bed, Hoblitt returns to ops. He announces he knows what the video is showing. The summit is coming apart. He puts the tape back in the VCR. "What we're seeing is a fissure propagating across the east flank like a zipper opening the mountain."

When they were out trying to fix the radio problem yesterday, Hoblitt says, he saw a widening base of ash. Maybe it was actually multiple vents opening. There could be a massive failure of the summit dome, leading to a catastrophic eruption.

I could be wrong, Hoblitt says. I'm tired and under a lot of stress.

"Yeah, but you've got common sense," says Colonel Grime.

Another geologist, Ed Wolfe, says Hoblitt's theory is one possible interpretation. But Hoblitt is convinced of it and says, "There are moments in history when tragedies occur because people won't face reality—I don't want this to be one of those moments."

He wants to inspect the edifice again. It's almost 6 A.M., and there may soon be enough visibility to fly. Grime says he is going, too. He says he wants to make sure the pilots don't get too close.

At 5:55 A.M., with the sky heavily clouded from the approaching typhoon, Pinatubo lets loose. The volcano has entered its cataclysmic eruption sequence. The nasty gray clouds go up in a great fountain, but only a short distance, and collapse onto the volcano's flanks. It's a Lamington-type eruption. The clouds sweep out from Pinatubo in every direction. The walls of ash quickly stretch almost three miles on all sides of the summit.[3] As it sweeps out from the summit, the churning ash cloud mixes with cool air, which heats and expands. At the same time, the larger rocks are settling out. Both processes cause the cloud to become more buoyant. Unless the group of Aetas that the repair group saw yesterday evacuated, they are dead. Colonel Freeman raises his camera. It has a wide-angle lens, but even from twelve miles away, the wall of the eruption spreads beyond the frame.[4] From side to side, it is now about 7.5 miles wide. The entire western horizon is boiling with churning clouds of ash. The edges become outlined in streaks of red and orange lightning bolts.

"Should we evacuate the base?" asks Anderegg.

Yes, says Hoblitt. Rick rushes outside to see if the ash clouds and

*Rick Hoblitt, Dick Anderegg, Andy Lockhart, and Dave Harlow in front of
Mount Pinatubo as it erupted at 5:55 A.M., June 15, 1991.*

pyroclastic flows, which now seem to cover the entire volcano, will reach
the base. Going from the meat-locker temperature of Dau to the humid
tropical air outside fogs Hoblitt's camera lens. He realizes the clouds
will stop short of the base. He gives Anderegg a thumbs-up. The boiling
cloud soon lifts off, rising on its own convection current. He says to
Anderegg, "Now it will go into a vertical eruption."

Harlow notes the total amount of seismic energy is not dropping off
this time. It's building.

Hoblitt then realizes that the Air Force is insisting the geologists evacuate, too. He hadn't thought the evacuation order would apply to them.

The geologists and senior officers are the last to leave Clark. As they
drive through the gates, Hoblitt is near tears. He is convinced the geologists shouldn't be leaving. The base wasn't overrun with pyroclastic
flows and the equipment can still function. At a predetermined staging
area, Hoblitt and other geologists say they want to return. The Air Force
relents. But if the geologists are going back, so will they.

"I think we just felt that there was important work to do there," recalls Wolfe. "There was still stuff to learn."

The squad of geologists, eighty-eight Air Force security cops, and about a dozen officers make it back to Clark before looters have had a chance to reach Dau.

While the geologists work to reestablish Dau, Colonel Anderegg decides to make a quick run to his house, which is uphill, past the Parade Field in the center of the base, to collect valuables left behind in the evacuation. Just as he drives past the west end of the Parade Field, he hears a senior cop withdrawing his troops from The Hill because a rolling cloud of ash is bearing down on them. Anderegg stops his car. Soon, he sees darkness fall ahead of him. Streetlights automatically begin turning on. Then, seconds later, the darkness grows thicker. He watches the streetlights disappear into the black cloud as if they had been turned off. Anderegg spins the car around and heads toward Dau. As he races downhill, he can't see anything in his rearview mirror but black. By the time he reaches the runways, he can see only to the end of his hood. His headlights glow only a few inches out. He slows his car and creeps along until he smacks into Dau.

About 10:30 A.M., the PIE record is a smear, as is Negron and BUGZ. CAB, the instrument on Clark that Andy got running just a few days ago, is the most distant instrument from the volcano so it has the greatest filtering, and it's showing high-frequency tremor.

Pinatubo erupts at 10:28 A.M., 11:42 A.M., and 12:52 P.M. Each eruption is followed by a complete blackout thirty minutes later.

"That's the thing that really surprised me about big eruptions," said Lockhart. "You hear people say, 'Well, it got completely dark.' 'It was dark as night.' I've always thought that's probably an exaggeration because of the way the thing forms. Basically, it forms a parasol, extending to the horizon. So chances are that it probably looks like night if you're looking straight up, but if you look out toward the horizon, almost assuredly you're going to see some little crack of light along the horizon, beyond the edge of the cloud. Well, no. It's just like the middle of a dark, stormy night. It's completely dark. You couldn't see the horizon. It's like a rainy, cloudy night when there's no starlight and no reflected light from the ground. I was astonished."

All of the eruptions since June 12 have shared this volcano's peculiar behavior pattern. Ewert's dome was blown out on June 12, and that eruption stopped when material from the collapsing column fell back into the vent, plugging it, allowing another dome to build and cork the vent. But each new plug was less and less effective in corking the vent. The plug just doesn't have time to crust over deeply enough to hold back the rising gas-charged magma. Now, with eruptions coming so frequently, the pipeline to the surface is nearly completely open for the rising stream of gas-charged magma.

Typhoon Yunya is now muscling its way through the ash and pumice. Anyone outside is quickly drenched with rain. But they are also hit with mudballs and pieces of pumice, small at first, like pellets, then marble-size, and finally the size of golf balls. For a volcano to be flinging pumice that size fifteen miles is awesome.

All anyone outside Dau hears is the clink of pumice hitting the building and the thunder of the mudflows rushing down the sides of Clark like powerful locomotives.

Pinatubo blows again at 1:16 P.M. Between eruptions, Pinatubo is hidden from Dau by the clouds of ash and the typhoon, which is now lashing the region with high winds and almost horizontal mud rains.

Everyone in Dau is clustered around the instruments. Following the 1:16 eruption, BUGZ's signal disappears. The interpretation is that it was taken out by a pyroclastic flow, which is not much of a concern because BUGZ was planted to the southwest side of summit. Because of the topography, big pyroclastic flows were expected to go down the west side.

About ten minutes later, BUGZ's signal is back on the screen. That's interesting, says Lockhart. We didn't lose the instrument, we lost the radio signal. Maybe when the pyroclastic flows move through one of the line-of-sight radio links, it blocks the signal. So maybe we can use these lost signals to figure out what's going on between us and the base of the volcano. So instead of sitting here blind, we might now have a way of seeing whether pyroclastic flows are coming at us.

Andy and the others are now glued to the computer screens tracing the seismic activity. Lockhart is angry at himself for not taking the risk to get UBO back up. UBO was just east of the summit. If that had failed, either intermittently or permanently, it would have provided a lot of information about the natural demolition going on outside their windows.

The next eruption, at 1:42 P.M., was the biggest of them all. At last, the gas-rich magma has direct access to the surface. An overpressurized lake of liquid rock, three miles wide, is jetting through an opening about one hundred yards wide. It's a runaway eruption.

Inside the throat of that volcano's jet, bubbles of gas are expanding in fresh liquid rock. Some of the stuff that falls from a volcano's column is this light, frothy rock, something of the consistency of thick Styrofoam. Sometimes, in this throat, the bubbles keep expanding until they burst and the pumice disintegrates. All that is left are relics of the bubble walls, ash. The magma is blowing itself to smithereens.

Seismographs are dying in every sector. When PIE's signal vanishes, it is certain that pyroclastic flows have vaulted the last natural barriers between the volcano and the base and are headed directly at Clark, another eight miles downhill.

There is no time to escape. There is no gate directly behind Dau. People would have to go outside, get into their vehicles, drive laterally to the gate, and leave. If the pyroclastic flows that took out PIE have enough energy, they can overwhelm the base in less than four minutes. There just isn't enough time to run.

Almost everyone instinctively moves away from the windows and into the back of the building. The few walls they put between themselves and the volcano might be enough. And they wait. Standing there, amid pallets of supplies, Andy Lockhart grabs a bag of popcorn and starts eating. Amazed, one of the PHIVOLCS geologists asks, "How can you eat popcorn at a time like this?"

"I always eat popcorn at this part of the movie," he says.

Hoblitt goes to the doorway to wait. General Studer and Colonel Grime join him. They concentrate on a row of red lights that line the runway. Those lights, a few hundred yards away, are all that can be seen. If those disappear, Rick thinks, that will be bad. The three men stand watching the lights. The rest of the world is entirely black. The lahars running in the rivers continue to rumble like subway trains as boulders are knocked together.

After a few minutes, it's apparent the threat from this pyroclastic flow has passed. What they don't know is that the pyroclastic flow made it to the wall at the top of The Hill.

People begin drifting back into the ops. They again discuss evacuation. Rick believes that lahars may have cut off their escape route and says it may be safer to stay. The others want to get the hell out. Pinatubo has gone Plinian. No one can see it, but it must be pumping up an enormous eruption column, and when it collapses, it will spray out more pyroclastic flows and some of those could reach the base. Someone says, since April 2 this volcano has only gone in one direction, worse. The next eruption could easily be worse than this one, and *those* pyroclastic flows could take out the base. Pumice pieces the size of Ping-Pong balls are falling. To throw something that size this far means the volcano is powered by astounding energy. This could be the most powerful eruption this century. As for science, only one instrument is still functioning, CAB, and it has been rail to rail for fifty minutes. And you can forget about any observations. When it isn't black as a coal mine outside, it's impenetrably gray from the typhoon. It's impossible to observe anything. It is time to go.

"There was nothing more we could do because the seismic network had been destroyed," says Wolfe. "Our monitoring tools were gone. At that point, there was absolutely nothing to be gained by our staying, so we left."

Everyone reaches for their backpacks and papers. Calmer heads grab cases of soda and MREs before running for the vehicles. Hoblitt gets into the embassy Suburban and finds the driver crossing himself. Hoblitt asks the driver if he knows where he's going. When he says no, Hoblitt gets out and looks for another car.

The caravan of official vehicles moves slowly. To keep the car ahead in sight, the car behind must follow within a few feet. In one, Ewert is gripping the wheel of a truck and saying over and over, "Shit. Shit. Shit." The roads are clogged with oxcarts, jeepneys, buses, and trucks, many of them overflowing with people hanging off them. The geologists are part of a wave of refugees.

It was at a point like this that Pliny the Younger guided his mother off the road, fearful that the enveloping darkness rushing at them would leave them blind and vulnerable in the road.

The car windows get caked with mud and the windshield wiper fluid is quickly used up. In his car, Hoblitt snaps open a can of cherry 7 UP, hangs out the passenger window, and pours it on the windshield. People in every car are doing the same thing. Hoblitt uses one of the empty soda cans to collect an ash sample from his hair and shirt pocket. At points, they pass individuals who are throwing buckets of water on windshields of the cars and trucks passing by. Only red taillights are visible ahead. They hope it is the vehicle they are supposed to be following. Brown rain and pumice smear windows. A group of women push their umbrellas against the wind and debris. Animals wander in and out of the headlights. Occasionally, light returns. It's a greenish, suffused light, like the eerie light that accompanies a tornado-packing thunderstorm in the Midwest. Says Ewert, "We're literally looking over our shoulders to see if there's an ash cloud descending on us." Then, it's black again.

"It was like running away in slow motion," says Ewert. "It was almost dreamlike because we couldn't go very fast. We're going maybe twenty miles an hour. I'd rather have been going eighty, but we couldn't go faster. We couldn't see. And there was lots of traffic."

Pinatubo's ash column now reaches twenty-two miles into the sky and is three hundred miles across.

At eight miles from Clark, the convoy is stopped. Gridlock; jeeps and water buffaloes. Talking to the embassy on the car phone, the scientists learn that Pinatubo's earthquakes are shaking Manila. Pumice is thumping against the metal of the vehicles. Inside, it sounds like a blizzard of hailstones. Thunder is cracking above, and a low subwaylike roar rises from behind. A woman comes up to Anderegg's car, holds out her baby, and pleads that he take the infant. Anderegg points in the direction of the agricultural college and says, "Go that way. You'll be okay."

Life at Subic is not much better. Quakes are bouncing the base. The winds from the typhoon have pushed ash clouds toward the Navy installation until the sky is as dark and the air as thick with ash, as Clark had been. A barrage of thunder and lightning bolts of green, blue, and red explode just overhead. Power is lost, water stops running, trees split. The roof on the high school begins to groan under the weight of the wet ash. Guards order everyone out of the collapsing building. Most people make it, but the nine-year-old daughter of an Air Force sergeant and her Filipino friend are killed.[5]

Buildings all around Pinatubo are groaning under thousands of pounds of wet ash, and they are being weakened by unceasing waves of earthquakes. Collapsing buildings are killing hundreds of people.

Stopped in the traffic, Dave Harlow uses the cell phone to call the U.S. embassy in Manila again. They report that Manila is still being rocked by big earthquakes. This is definitely not good. To feel earthquakes sixty miles from a volcano is not normal. These are trouble. Up until now, quakes have rarely been felt even on the base. The list of possibilities is short. In fact, probably only one thing could be causing such big earthquakes. As the volcano pumps out vast amounts of material from underground, large portions of the volcano's roof must be falling into the collapsing magma chamber. A caldera might be forming.

Three hours after leaving Clark, the caravan of geologists and Air Force personnel have traveled only ten miles and reached the Pampanga Agricultural College at the base of Mount Arayat. The Air Force establishes a command post in a second-floor classroom. Like the rest of central Luzon, the college is without power, and the command center glows green from fluorescent light sticks. When Hoblitt arrives, he is told quakes are continuing in Manila. He didn't need to be told. Even here, the world is rocking. These are the biggest earthquakes the geologists have felt in the Philippines.

When the scientists arrive at the agricultural college, the Air Force informs them that imagery of the volcano shows ash clouds rising along a line about three miles long.[6] Hoblitt is shocked. If that ash is coming from an extended series of vents, it may mean that a large caldera-forming eruption is in progress. This is consistent with the big seismicity. Hoblitt informs PHIVOLCS of his concerns and suggests extending the hazard boundary to eighteen miles.

The danger from a caldera-forming eruption is from huge pyroclastic flows. They emerge from a circle of cracks that form when the roof rocks sink into the magma chamber. Pyroclastic flows are pumped out like from a bellows. The collapse can be three to six miles across, or more.

Hoblitt digs into the collection of publications he has with him to see what he can find about the reach of pyroclastic flows from caldera eruptions. He sits in the dark, reading the book *Volcanic Successions* by holding a MagLite in his mouth. He decides there is no danger from pyroclastic flows at their current location, which he has been told is thirty-four miles from the volcano. (The site is actually twenty-one miles from Pinatubo.)

He is checking out the description of Crater Lake. Over sixty-eight hundred years ago, a volcano very much like St. Helens in what is now southwestern Oregon exploded in an enormous eruption. So much material was ejected that a caldera was formed, which over thousands of years trapped water and became Crater Lake. The caldera's formation sent pyroclastic flows thirty-seven miles.

There is really nothing for the geologists to do now but wait. They are without instruments. If they knew they were at twenty-one miles, they couldn't have gone any farther anyway because the roads are clogged. Nothing is moving.

The geologists pick an empty classroom on the second floor. They find sleeping bags, which were stored there weeks ago by Colonel Murphy. Lying on the floor, the geologists can feel the earthquakes rolling in. They can even feel the two waves of a quake, as Frank Perret did with his teeth.

A big jolt bounces the building, and everyone bails out. Many of the Air Force people were around for last year's 7.8 quake, and they saw a lot of buildings collapse. One of the geologists does a quick structural evaluation of the building, which seems sound. Most important, the roof is not heavy. It's made of corrugated metal. This will withstand big quakes, says the geologists. So the VDAP scientists return to the classroom, lie down, and are rocked to sleep. They are filthy and exhausted, but for now, their job is done.

Outside, the bar girls are prowling the military vehicles looking for business. The barbers have also relocated, and senior officers line up to take their turn.

Pinatubo's runaway eruption would keep jetting for nine hours.

While the geologists sleep, Pinatubo does form a caldera, but its pyroclastic flows are not large enough to reach Angeles. Amazingly, few lives are lost directly through the eruption or the associated pyroclastic flows in what turns out to be the second-biggest eruption of the century. Despite the timely warnings, hundreds of lives are lost as buildings, weakened by earthquakes and weighted down by the wet pumice, collapse. The wet ash was so heavy twenty miles away at Cubi Point Naval Air Station, south of Subic Bay, that when rain-soaked ash piled on the tail of a parked DC-10, it rolled nose up.[7]

The Air Force returned to Clark the next day. Actually, they had never really left. When the first officers made their way back to Clark, they were greeted by Air Force security. The cops were assigned to direct traffic during the evacuation and no one had told them to leave. Rather than abandon their posts, they continued to direct traffic throughout the eruption.

Inside the wall, the sight that greeted the base commander and the other Air Force officers nearly brought them to tears. Their slice of

Aerial view of section of Clark Air Base after the eruption of Mount Pinatubo. The gymnasium and other buildings collapsed under the load of tephra from the eruption. (PHOTO: WILLIE SCOTT)

America had been wrecked. Everything was covered with six to nine inches of compressed ash. Jet hangars, which had housed the security force just before evacuation, were flattened. Supply houses and gymnasiums were crushed. The towering acacia trees lay in heaps. At least one mudflow had ripped through the base. In all, one hundred buildings had been destroyed. The base had sustained $300 million in damage.[8] The base negotiations were essentially moot. On July 17, 1991, it was made official: the United States would abandon Clark.

Outside the base, geologists found pumice rocks the size of lemons near the western wall. They also found that one pyroclastic flow had come to within four hundred yards of The Hill. Crow Valley, the hottest mock-war zone in the Pacific with its wicker buildings, was nearly filled with flows. At places, pyroclastic flows were six hundred feet thick.[9] And the top of Pinatubo was gone. In its place was a caldera about 1.5 miles wide.[10]

Pinatubo had threatened more than 1 million people.[11] Over eighty thousand people within Pinatubo's hazard zones had been evacuated.[12] Perhaps as many as twenty thousand escaped death.[13] Slightly more than two hundred people died, most of those from roof collapses[14] and a few dozen in pyroclastic flows and mudflows.[15] In other words, less than one-quarter of one percent of those at risk had died during the eruption. This was not Ruiz. Despite a cost of $1.5 million, the monitoring operation was estimated to have saved at least $375 million in damage and probably much more.

As Hoblitt had estimated, Pinatubo had ejected about five cubic kilometers of magma.[16] Ten times the size of St. Helens, it was the second-biggest eruption of the century. But it was still one of Pinatubo's modest outbursts.

In the months, even years, following the eruption, that corner of the Philippines would be repeatedly brutalized during the rainy season. As they had done for tens of thousands of years, storms would pass over the ash flows and thick pyroclastic flows. The rains would mix with the debris, and soon broad lahars were racing downhill, often across farmlands and sometimes through villages. Hundreds of people would continue to die in these annual mud floods.

In an after-action memo that dissected the problems the VDAP team faced at Pinatubo, Newhall blisteringly criticized the skeptics, including the director of the local US AID program.

"As a direct result of this skepticism, we very nearly couldn't get the job done," he wrote. He said that because of an inadequate response, the Survey team had been delayed in reaching the Philippines. US AID funds were gone within days of arriving. Without funds, it was impossible to charter a plane for COSPEC measurements. Senior officers with the Air Force limited helicopter time. But Newhall's harshest criticism was of his official "host," US AID.

"Skepticism from the US AID country director, our nominal 'host,' also made our job of convincing various officials more difficult. In effect, official skepticism within the [U.S. government] nearly killed a large number of people."[17]

Luck played its role. Some volcanoes have rushed from first quakes to full destruction in less than a week. Pinatubo took a full two months. While that was not much time, it was enough for geologists to deploy their portable observatory, begin to get a feel for the volcano from the monitoring, and establish their credibility with those at risk.

Airlines suffered dramatic losses. Commercial jets had several encounters with Pinatubo's ash. The pilot of a Saudi 747 ignored ash warnings and landed at Manila during the eruption. He inflicted millions of dollars of damage on his plane. Across Southeast Asia, 747s running high and hot inhaled tons of volcanic ash and lost engines. A few lost all four engines and plummeted thousands of feet. But all were able to restart before crashing. Two dozen planes on the ground were damaged by the eruption. Another sixteen encountered ash in the air. The bill for each aircraft/ash encounter totaled over $100 million. It was obvious that Newhall's aircraft/ash conference should have been well attended.

Pinatubo demonstrated better than any other volcano how valuable geological smoke jumpers could be. The Pinatubo eruption marked the beginning of the truly mobile volcano response team. It also opened the door to the rest of the world. Pinatubo was VDAP's first volcano outside of Latin America; now they branched out to the rest of the world. The VDAP team also learned from Pinatubo. They immediately acquired a portable weather radar so that they would not be blinded by the dark or storms.

Future teams, which often included Rick Hoblitt, Andy Lockhart, and John Ewert, would go on to provide important guidance to threatening volcanoes from Papua New Guinea to the Caribbean.

The Three Musketeers went on to have enormously productive careers within the Survey. Don Swanson became the scientist-in-charge at Hawaiian Volcano Observatory. By the late 1990s, he was doing a stratographic analysis of Hawaiian volcanoes, just as Crandell and Mullineaux had done of the Cascades.

In the years after Pinatubo, the cultural division within the Survey's volcano program between meatballs and coneheads grew more bitter. Publications were what counted toward career advancement in the Survey, not saving lives. Despite noble words to the contrary, it's still largely true today. Moreover, Survey officials, mostly HVO grads themselves, continue to claim that rotating geologists through HVO is essential for maintaining an expert crisis-response team. But, in fact, at Pinatubo there was only one HVO grad, CVO chief Ed Wolfe.

So Dave Harlow, the seismologist with the most experience on explosive volcanoes, resigned. After thirteen years in the same pay grade, the king of the meatballs left the Survey. Pinatubo made it easy, he would say years later. For nearly twenty-five years after leaving Vietnam, the ex-Marine had not been able to find peace with that part of his life. The human violence and hollow homecoming fed a smoldering anger and relentless guilt. The Philippines—from the rice paddies to the military chain of command—brought back memories of Vietnam, which were magnified by the stress of the crisis. But Pinatubo had turned out well.

Once back home in Menlo Park, Harlow was surprised to notice a change in himself. As he watched a television program about Vietnam, it suddenly occurred to him that he wasn't feeling the same steam he often did when anything touched on the subject.

"I felt as though my karma had kind of straightened out about Vietnam," he said. "It's a combination of feeling as though I made, in this particular case, a real contribution to saving a lot of lives and doing a good thing."

After leaving the Survey, he began working on the streets of Menlo Park with high-risk teenagers.

Field Notes

Whurhen Hoblitt landed in Portland, Oregon, he wrote the last words in his Pinatubo field notebook: "It's over." Of course, it wasn't. Norm Banks's concept of a quick-reaction team of volcanologists who can deploy their portable volcano observatory anywhere in the world had proved to be a success. So, Lockhart, Ewert, Newhall, Hoblitt, and the many others would continue to fly off to attend to other loaded volcanoes, year after year.

When Hoblitt returned to his basement office at the Cascades Volcano Observatory, he felt as though he never wanted to travel again. But as ever, his friend Dan Miller had other plans. Miller had arranged with the Soviet Union for the first official visit by American scientists to the volcano most like St. Helens, Bezymianny. Just prior to May 18, 1980, the geologists had argued over the 1959 paper on a Bezymianny eruption. The central feature of that 1956 eruption was a lateral blast. Back in 1980, many of the geologists at St. Helens were not confident about the work of the Soviet scientist who had studied the blast and written his paper years after the eruption. Moreover, a lateral blast was almost unheard of, and a single paper by a Russian author was generally dismissed. It had been a mistake. Now, Miller had arranged through Soviet scientists he had met at an international meeting in Japan to visit Bezymianny. Hoblitt took some convincing, but as ever, Miller's influ-

ence over him was too great. So, a few months after the Pinatubo eruption, the two geologists headed off to another eruption, this time on the
Kamchatka Peninsula.

For three weeks, they lived in canvas tents on the flows of Bezymianny. The volcano was eerily similar to St. Helens. It, too, had a debris
avalanche spread before it, and a ramp that led to a horseshoe-shaped
amphitheater. But the dome growing inside the crater had nearly filled
the gaping wound. And it was still growing, with a constant series of
dome-building eruptions. Hoblitt and Miller worked long hours as, once
again, a foreign volcano was proving to be a scientific treasure.

The living conditions on the flows were primitive, but Dan Miller
had a way of making the best of any situation.

He had packed one of the big VDAP trunks full of exotic foods. They
had cans of smoked oysters and salamis and smoked clams and cheeses
and nuts and all kinds of candies and other things that Soviet scientists
couldn't get at the time. Miller had also brought two bottles of rum and
Hoblitt had brought a bottle of rum and a bottle of tequila. When it
came to liquor, the Soviets were not to be outdone. They had rounded
up twenty bottles of vodka and a case of beer. All of it, the food and
drink, produced a nice little cocktail hour after darkness fell and work
ended for the day.

Following an old Russian tradition, before each drink someone had
to give a toast. The Soviets, of course, were impressive toasters and gave
long, thoughtful, and often moving speeches.

"Some of them about brought tears to your eyes," says Miller. "They
really say things from the heart."

For example, one of the women seismologists had agreed to leave
her children for three weeks and cook for the Americans. She didn't
speak English, but she spent one whole day, while the others were out in
the field, preparing her toast, using a Russian-English dictionary to
translate every word. She gave her toast in halting English, but the
meaning was clear. She said that she had not wanted to leave her children to cook for the Americans, but she was now glad she had because
she had been rewarded with these new friends. It was, she said, the
most meaningful thing that had happened to her in years.

They waved their tin cups of vodka while embers jumped off the
campfire like fireflies. Every once in a while, they stopped toasting to
watch incandescent hot avalanches tumbling off the dome. And the

cultural divisions between the two groups eroded until they were just a group of people who shared a love of volcanoes.

At first, the two American scientists had been awkward saying anything to an audience without a slide projector. But as the drinks warmed their spirits and loosened their reserve, they began perfecting the art of Russian toasting. No one now remembers exactly what Miller and Hoblitt said, and perhaps the details aren't really important. But on that plain of volcanic debris, with Bezymianny lava flowing in the background—with Pinatubo behind it, and Ruiz and St. Helens and Pelée and Krakatau and Vesuvius behind it—the little group of scientists sitting around a campfire in the middle of a vast, desolate stretch of the Kamchatka Peninsula raised their tin cups to Harry Glicken, to the Kraffts, and to Dave Johnson. To them all.

Notes

INTRODUCTION

1. Richard Fisher, Grant Heiken, and Jeffrey Hulen, *Volcanoes: Crucibles of Change* (Princeton, N.J.: Princeton University Press, 1997), 93.
2. Kilauea does have an impressive explosive history. It exploded in 1924, killing one person. In 1790, an explosive eruption killed at least 80, and possibly several hundred, people. But in the experience of the Survey scientists who trained on Kilauea and Mauna Loa, and went on to work at St. Helens, the Hawaiian volcanoes produced deceptively gentle eruptions. Personal communication from Don Swanson.

CHAPTER 1: HOBLITT'S FLOATING ISLAND, SUMMER 1979

1. R. I. Tilling, "History and Major Accomplishments of the Volcanic Hazards Program" (November 1989, unpublished).
2. Fisher, Heiken, and Hulen, *Volcanoes*, 131.
3. R. Crandell and D. Mullineaux, *Environmental Geology*, vol. 1 (New York: Springer-Verlag, 1975) 23.
4. Blue Book, C2.
5. Robert Tilling, "Volcanic Hazards and Their Mitigation: Progress and Problems," *Reviews of Geophysics*, May 2, 1989.

CHAPTER 2: DISBELIEF

1. Steve Malone, "Preliminary Seismic Response Log."

2. T. Saarinen and J. Sell, *Warning and Response to the Mount St. Helens Eruption* (State University of New York Press, 1985), 40.

3. P. Sheets and D. Grayson, *Volcanic Activity and Human Ecology* (Academic Press, 1979), 225. The authors found evidence that the local economy was already in decline by the time the volcano emergency struck. However, at the time, the decline was attributed to the closures.

4. D. Frank, M. Meier, and D. Swanson, "Assessment of Increased Thermal Activity at Mount Baker, Washington, March 1975–1976," USGS paper 1022-A.

5. Bob Christiansen, memo to Bob Tilling, chief, Office of Geochemistry and Geophysics, February 25, 1980.

6. U.S. Forest Service, "On the Mountain's Brink: A Forest Service History of the 1980 Mount St. Helens Volcanic Emergency," 13.

7. *Oregon Statesman* newspaper, March 27, 1980.

8. Jerry Brown, "Role of the U.S. Forest Service at Mount St. Helens," Special Publication 63, proceedings of a workshop on volcanic hazards in California, December 3–4, 1981 (California Department of Conservation, Division of Mines and Geology), iii–74.

9. J. Sorensen, "Emergency Response to Mount St. Helens' Eruption: March 20 to April 10, 1980" (Energy Division, Oak Ridge National Laboratory), 71.

10. D. Crandell, D. Mullineaux, C. Miller, "Volcanic-Hazard Studies in the Cascade Range of the Western United States," in *Volcanic Activity and Human Ecology* (Academic Press, 1979), 195.

11. J. Sorensen, "Emergency Response," 22.

12. USGS professional paper 1249, "The Volcanic Eruptions of 1980 at Mount St. Helens: The First 100 Days," 18.

13. *The Fire Below Us: Remembering Mount St. Helens.* Film documentary by Global Net Productions.

14. Tape 1, side 1.

15. "Volcanic and Seismic Activity at Mount St. Helens," *USGS Monthly Report,* March–April 1980, 2.

16. "Report of Geologists' Flight," USGS memorandum, March 27, 1980.

17. Richard Fiske, "Volcanologists, Journalists, and the Concerned Public," in *Explosive Volcanism: Inception, Evolution and Hazards* (National Academy of Press, 1984), 170.

18. R. Kerr, "Mount St. Helens: An Unpredictable Foe," *Science Magazine,* June 27, 1980, 1448.

19. Blue Book, C2.

20. Saarinen and Sell, *Warning and Response,* 34.

21. D. Crandell, "This Is My Life" (unpublished autobiography), 94.

22. Saarinen and Sell, *Warning and Response,* 50.

23. USFS, "On the Mountain's Brink," 15.

24. Saarinen and Sell, *Warning and Response,* 187.

25. USFS, "On the Mountain's Brink," 17.

26. USGS paper 1249, p. 23.

27. Greene, Perry, and Lindell, *The March 1980 Eruptions of Mt. St. Helens: Citizen Preceptions of Volcano Threat* (Battelle Human Affairs Center), 49.

28. Rob Haesler, "People Cool Near Volcano," *San Francisco Chronicle*, March 29, 1980, 1.

29. John O'Ryan, "I'm Not Leaving Until the Mountain Has Blown," *Seattle Post-Intelligencer*, March 27, 1980, A6.

30. United Press International, March 28, 1980.

CHAPTER 3: THE MUSKETEERS

1. "Volcano Boils, Scientists Puzzle," *Merced* (Calif.) *Sun-Star*, March 29, 1980, 10.

2. R. Kerr, "Mount St. Helens," 1446.

3. *USGS Monthly Report*, March–April 1980, 5.

4. Ibid., 7.

5. USGS paper 1249, p. 22.

6. Fisher, Heiken, and Hulen, *Volcanoes*, 48.

7. USGS professional paper 1250, "The 1980 Eruptions of Mount St. Helens, Washington" (1981), 190.

8. Saarinen and Sell, *Warning and Response*, 44.

9. *USGS Monthly Report*, March–April 1980, 12.

10. *Northwest Magazine*, NW25, July 13, 1980.

11. S. Hobart, "Volcanologist Doesn't Expect Lava Eruption," *Portland Oregonian*, April 6, 1980.

12. U.S. Department of Interior, Geological Survey, National Center press release, "USGS Director to Observe Mount St. Helens Volcano," April 9, 1980.

CHAPTER 4: THE BULGE

1. Hoblitt's notes, April 14, 1980, 12:27 P.M.

2. Fisher, Heiken, and Hulen, *Volcanoes*, 279.

3. "Scientists Find St. Helens a Typical Volcano," *Gazette-Times* (Corvallis, Oreg.), April 17, 1980, 15.

4. Hoblitt E-mail and Polaroid photograph.

5. *USGS Monthly Report*, March–April 1980, 3.

6. USGS paper 1250, p. 154.

7. *USGS Monthly Report*, March–April 1980, 13.

8. "Bulge Newest Hazard at Mount St. Helens Volcano," USGS hazard announcement, April 30, 1980.

9. USGS paper 1249, p. 37.

10. "Ray Further Restricts Peak Access," *Seattle Times*, May 1, 1980, B9.

11. "Quakes Don't Scare Residents of St. Helens Area," Associated Press/*Gazette-Times* (Corvallis, Oreg.), April 14, 1980, 14.

12. USGS paper 1249, p. 35

13. Ibid., 37.

14. Ibid.

15. "Life with Volcano Problematic," *Portland Oregonian*, April 13, 1980.

16. "Some Expect Lava; Others Not So Sure," *Sunday Columbian,* May 11, 1980, 1.
17. USGS paper 1249, p. 38.
18. *Times* staff, "Unlock the Gate and Step Aside—St. Helens' Cabin Owners to Defy Roadblock," *Seattle Times,* May 17, 1980, 1.
19. Saarinen and Sell, *Warning and Response,* 72.

FIELD NOTES: MAY 17, 1980/MINDY BRUGMAN

1. Fisher, Heiken, and Hulen, *Volcanoes,* 26
2. USGS paper 1249, p. 42.
3. Ibid.
4. Saarinen and Sell, *Warning and Response,* 188.
5. USFS, "On the Mountain's Brink," 28.
6. "Hawaii Volcanologists Help Keep Tabs on St. Helens," *Honolulu Sunday Star-Bulletin & Advertiser,* May 18, 1980.

CHAPTER 5: SWANSON

1. USGS paper 1250, p. 809.
2. USGS paper 1249, p. 56.
3. USGS paper 1250, p. 809.
4. USFS, "On the Mountain's Brink," 37.
5. USGS paper 1249, p. 60.
6. USGS paper 1250, p. 809.
7. Ibid., 492.
8. Ibid., 809.
9. Ibid., 492.
10. H. Glicken and M. Brugman, "Eyewitness Report of Harry Glicken," from notes and interview, USGS memo, May 18, 1980.
11. USGS paper 1250, p. 580.
12. USGS paper 1249, p. 59.
13. USGS paper 1250, p. 580.
14. Ibid., 331.
15. S. Wagner, "Ray Asks Disaster Status for Washington," *The Oregonian,* May 21, 1980, 1.
16. USGS paper 1250, p. 672.
17. Ibid., 667.
18. Ibid., 715.
19. USFS, *On the Mountain's Brink,* 36.
20. USGS paper 1249, p. 69.
21. Fitzgerald had been working on his Ph.D. in geology, which his friends finished for him. The degree was awarded posthumously.

FIELD NOTES: MAY 18, 1980/THE ST. HELENS OBSERVERS

1. USGS paper 1250, p. 53.
2. Ibid, 349.
3. Ibid.
4. USGS paper 1249, p. 45.
5. Ibid., 47.
6. U.S. Forest Service, "Reflections—St. Helens 10 Year Later," 4.
7. USGS paper 1250, p. 131.
8. Ibid., 344.
9. It was recorded by a local ham radio operator who weeks after the blast gave a copy to Dan Miller.
10. J. Moore and C. Rice, *Chronology and Character of the May 18, 1980, Explosive Eruptions of Mount St. Helens* (National Academy of Sciences), 133.
11. USGS paper 1250, p. 2.
12. Pringle, P. *Roadside Geology of Mount St. Helens National Volcanic Monument and Vicinity.* Washington Department of Natural Resources, 1993, 31.
13. Forest Service poster "The whole west side—northwest side is sliding down."
14. USGS paper 1250, p. 393.
15. USGS paper 1249, p. 56.
16. USGS paper 1250, p. 351.
17. USGS eyewitness reports: Francisco Valenzuela. Recorded May 22 at Cougar evacuation center.
18. Kran Kilpatrick, quoted in U.S. Forest Service, "On the Mountain's Brink," 31.
19. USFS, "On the Mountain's Brink," 31.
20. Pringle, *Roadside Geology,* 31.
21. USGS paper 1250, p. 395.
22. USFS, "On the Mountain's Brink," 36.
23. USGS, paper 1249, p. 59.
24. USGS survivor interviews: Capt. Joseph Mathes by Richard B. Waitt, Jr., July 2, 1980.
25. R. Decker and B. Decker, *Scientific American,* March 1981, 84.
26. USGS paper 1249, p. 56.
27. USGS paper 1250, p. 809.
28. Ibid., 347.
29. USGS paper 1249, p. 45.
30. Pringle, *Roadside Geology,* 31.
31. USGS paper 1250, p. 449.
32. Ibid., 450.
33. Ibid.
34. Ibid., 470.
35. USGS paper 1249, p. 63.
36. USGS paper 1250, p. 483.
37. Ibid., 460.
38. Ibid., 463.

39. Ibid., 462.
40. Ibid., 479.
41. Ibid., 470.
42. Ibid.
43. Ibid.
44. Ibid.
45. USGS paper 1249, p. 64.
46. Ibid., 61.
47. USGS paper 1250, p. 482.
48. Ibid., 480.
49. USGS paper 1249, p. 65.
50. Ibid., 66.
51. Pringle, *Roadside Geology*, 32.
52. Ibid.
53. Fisher, Heiken, and Hulen, *Volcanoes*, 118.
54. Pringle, *Roadside Geology*, 33.
55. USGS paper 1250, p. 488.
56. USGS paper 1249, p. 60.
57. USGS paper 1250, p. 488.
58. D. Mullineaux statement, hearing of the Senate Commerce Committee, June 13, 1980.

CHAPTER 6: THE VOLCANO LAB: ST. HELENS AFTER THE BLAST

1. Rob Carson, *Mount St. Helens: The Eruption and Recovery* (Sasquatch Books, 1990), 56.
2. USGS paper 1249, p. 103.
3. Ibid., 76.
4. Ibid., 110.
5. Ibid., 66–99.
6. Carson, *Mount St. Helens*, 53.
7. USGS paper 1249, p. 90.
8. Ibid., 77.
9. Ibid., 78.
10. Ibid., 89.
11. Carson, *Mount St. Helens*, 74.
12. Yoshiaki Ida and Barry Voight, quoting Swanson, "Introduction: Models of Magmatic Processes and Volcanic Eruptions," *Journal of Volcanology and Geothermal Research* 66 (1995): x.
13. USGS paper 1249, p. 99.
14. Ibid., 97.
15. Ibid., 104.
16. Ibid., 106.
17. Robert Decker, personal communication.
18. Don Swanson, quoted in the *Journal of Volcanology and Geothermal Research* 66 (1995): xii.
19. Carson, *Mount St. Helens*, 71.

20. D. A. Swanson et al., "Predicting Eruptions at Mount St. Helens," *Science*, September 30, 1983, 1369.

21. Ibid., 1373.

CHAPTER 7: MAMMOTH LAKES: BETWEEN A ROCK AND A HARD PLACE

1. William Spangle and Associates, Inc., "Living with a Volcanic Threat: Response to Volcanic Hazards, Long Valley, California" (Consolidated Publications, Inc., 1987), 9.

2. Associated Press, "Area to be Notified of Possible Volcanic Activity," May 25, 1980.

3. David Hill, "Science, Geologic Hazards, and the Public in a Large, Restless Caldera," *Seismological Research Letters*, September/October 1998, 401.

4. John Carey, "The Earth Rocks and Rolls," *Newsweek*, January 24, 1983, 71.

5. Disaster Relief Act of 1974 (public law 93-288; 88 Stat. 143). Amended in 1976 to order the director of the USGS to issue "disaster warnings for an earthquake, volcanic eruption, landslide, mudslide or other geological catastrophe." The director is ordered to "provide technical assistance to State and local governments to insure that timely and effective disaster warnings are provided."

6. United Press International, regional news, June 11, 1982, A.M. cycle.

7. Survey scientist Robert Cockerham, quoted by UPI in "Scientists Say Eruption Could Trap Residents," July 8, 1982.

8. According to Bill Taylor, who was a real estate broker, the Mammoth real estate market, which was "on fire" from 1978 to 1981, had begun to slow as potential buyers realized Mammoth was a "very artificially inflated market" (*Mammoth Times*, August 11, 1988). The Suvey warning was issued as the market was collapsing.

9. Sandra Blakeslee, "Volcano Warning Brings Economic Woe to Coast Resort Area," August 12, 1984, 20. *New York Times*

10. Spangle and Associates, "Living with a Volcanic Threat," 9.

11. Hill, "Science, Geologic Hazards, and the Public," 403.

12. John Nobel Wilford, "In Sierra Nevada, Ominous Tremors Could Mean Another Mount St. Helens," *New York Times*, July 20, 1982, 17.

13. Resolution 82-121, Mono County, July 13, 1982.

14. Carey, "Earth Rocks and Rolls," 71.

15. Charles Petit, "Scary View: Mammoth Lakes Eruption by Late 1984," *San Francisco Chronicle*, October 18, 1983, 5.

16. Hill, "Science, Geologic Hazards, and the Public," 401.

17. Spangle and Associates, "Living with a Volcanic Threat," 13.

18. Ibid, 9.

19. Hill, "Science, Geologic Hazards, and the Public," 402.

20. George Alexander, "Volcanic Hazard Downgraded at Mammoth—At Least on Paper," *Los Angeles Times*, January 7, 1985, 3.

21. Martin Dubin, letter to Secretary of the Interior James Watt, November 3, 1982.
22. William Oscar Johnson, "A Man and His Mountain," *Sports Illustrated*, February 25, 1985, 70.
23. Hill, "Science, Geologic Hazards, and the Public," 404.

CHAPTER 8: THE VOLCANO ZOO

1. G. A. M. Taylor, *The 1951 Eruption of Mount Lamington, Papua*, BMR (Bureau of Mineral Resources, Geology, and Geophysics) Bulletin no. 38 (Australian Government Publishing Service, first edition 1958; second edition 1983).
2. I. Suryo and M. C. G. Clarke, "The Occurrence and Mitigation of Volcanic Hazards in Indonesia as Exemplifed at the Mount Merapi, Mount Kelut, and Mount Galunggung Volcanoes," *Q. J. Eng. Geol. 18* (London, 1985): 79.
3. Ibid., 80.
4. Ibid., 81.
5. Maurice Krafft, *Volcanoes: Fire from the Earth* (Harry N. Abrams Publishers, 1993), 117.
6. Robert Decker and Barbara Decker, *Volcanoes*, 3rd ed. (W. H. Freeman and Company, 1997), 290.
7. Fisher, Heiken, and Hulen, *Volcanoes*, 167.
8. Decker and Decker, *Volcanoes*, 290.
9. Suryo and Clarke, "Occurrence and Mitigation of Volcanic Hazards," 85.
10. Ibid., 83.

CHAPTER 9: AFTER ARMERO

1. Barry Voight, "The 1985 Nevado del Ruiz Volcano Catastrophe: Anatomy and Retrospection," *Journal of Volcanology and Geothermal Research* 44 (1990): 349–86.
2. Associated Press, "20,000 Feared Dead in Mud Slide," November 15, 1985.
3. Joseph Treaster, "15,000 Feared Dead in Colombia," *New York Times*, November 15, 1985, 1.
4. Associated Press, "20,000 Feared Dead."
5. Associated Press, "Chronology of Nevado del Ruiz Eruption," November 18, 1985.
6. Treaster, "15,000 Feared Dead," 1.
7. Associated Press, "20,000 Feared Dead."
8. Treaster, "15,000 Feared Dead," 1
9. Voight's paper was published by Penn State in 1987 and widely circulated within the USGS the following year.
10. Voight, "The 1985 Nevado del Ruiz Volcano Catastrophe," 380.
11. Jon Krakauer, "Geologists Worry about the Dangers of Living under the Volcano," *Smithsonian*, July 1996, 34.

12. Crandell, Mullineaux, and Miller, "Volcanic-Hazard Studies," 205.
13. Krakauer, "Geologists Worry about the Dangers," 39.
14. National Research Council, *Mount Rainier, Active Cascade Volcano* (Washington, D.C.: National Academy Press, 1994), 3.
15. Krakauer, "Geologists Worry about the Dangers," 36.

FIELD NOTES: DECEMBER 15, 1989/REDOUBT, ALASKA

1. Thomas J. Casadevall, "The 1989–1990 Eruption of Redoubt Volcano, Alaska: Impacts on Aircraft Operations," *Journal of Volcanology and Geothermal Research* (1994): 301–16.

CHAPTER 10: TRAINED DECISION MAKERS

1. Christopher Newhall and Raymundo Punongbayan, *Fire and Mud: Eruptions and Lahars of Mount Pinatubo, Philippines* (Philippine Institute of Volcanology and Seismology and University of Washington Press, 1996), 1.
2. Ibid., 191.
3. Ibid., 3.
4. Ibid.
5. Alberto Garcia-Saba, "The Hand of God?" *Time*, July 1, 1991.
6. Richard Anderegg, *Pacific Jewel* (USAF, 2000).
7. Newhall and Punongbayan, *Fire and Mud*, 75.
8. Anderegg, *Pacific Jewel.*
9. Ibid.
10. Ibid.
11. C. G. Newhall and R. S. Punongbayan, "Final Report of Collaboration," VDAP administrative report, July 1997, 17.

CHAPTER 11: THEY'LL THINK YOU'RE A HERO

1. In fact, a small, tragic eruption at Galeras did kill six volcanologists who were on a field trip inside the crater.
2. Newhall and Punongbayan, *Fire and Mud*, 73.
3. Fisher, Heiken, and Hulen, *Volcanoes*, 98.
4. Newhall and Punongbayan, *Fire and Mud*, 91.
5. Anderegg, *Pacific Jewel.*
6. Newhall and Punongbayan, *Fire and Mud*, 91.
7. Newhall and Punongbayan, *Fire and Mud*, 91.
8. Edward W. Wolfe, "The 1991 Eruptions of Mount Pinatubo, Philippines," *Earthquakes & Volcanoes* (USGS) 23, no. 1, (1992): 18.
9. Fisher, Heiken, and Hulen, *Volcanoes*, 69–70.
10. Ibid., 71.

CHAPTER 12: ERUPTION

1. Anderegg, *Pacific Jewel.*
2. Newhall and Punongbayan, *Fire and Mud,* 11.
3. Wolfe, "The 1991 Eruptions," 20.
4. Anderegg, *Pacific Jewel.*
5. Anderegg, *Pacific Jewel.*
6. This information later proved to be wrong.
7. Newhall and Punongbayan, *Fire and Mud,* 1080.
8. Peter Grier, "Last Days at Clark," *Air Force Magazine* (Air Force Association), February 1992, 56.
9. USGS fact sheet 115-97.
10. Anderegg, *Pacific Jewel.*
11. Newhall and Punongbayan, *Fire and Mud,* 1.
12. Richard Kerr, "A Job Well Done at Pinatubo Volcano," *Science,* August 2, 1991, 314.
13. Newhall and Punongbayan, *Fire and Mud,* 67.
14. Ibid., 15.
15. Ibid., 81.
16. Ibid., 1.
17. Newhall and Punongbayan, "Final Report of Collaboration," 17.

Glossary

ASH. Fine particles formed in explosive eruptions.

CALDERA. A gigantic depression in the earth formed when a volcano collapses into its underground magma chamber.

DOME. A mound of lava extruded onto the surface of a volcano.

FAULT. A break in the earth's crust.

LAHAR. A volcanic mudflow.

LATERAL BLAST. A hot mixture of gas and debris that moves at great speed over the ground.

LAVA. Molten rock on the surface, as distinct from magma, which is molten rock beneath the surface.

MAGMA. Molten rock beneath the surface, as distinct from lava, which is molten rock on the surface.

MAGMA CHAMBER. A reservoir of molten rock beneath a volcano.

MUDFLOW. A thick mixture of water and debris that can be very fast-moving.

PUMICE. Styrofoam-like material ejected during an explosive eruption.

PYROCLASTIC FLOW. Fluid cloud of hot gases and volcanic ash that can move at hurricane speeds.

VENT. Opening in a volcano through which volcano material is ejected.

Acknowledgments

After more than two decades of exciting and rewarding group journalism at *Time* magazine, I wanted to produce a story that was entirely mine. It took me two years to finish this book; by then I realized that no work of nonfiction is entirely the author's. This has been a collaboration with those who gave me their time for interviews, who opened storage bins to dig out papers and journals and complex memories, and who gave me their honest judgments as this book moved toward completion. I am deeply grateful for all their contributions to this book.

I also want to thank my teachers—Tom Johnson, Albert Low, and Janet Richardson.

My agent, Kris Dahl, had faith in the project even as we accumulated two dozen rejections. And my editor at St. Martin's, Ruth Cavin, was everything a first-time book writer would hope to find—a consummate editor, a stern taskmaster, and a warm person who always had time to listen to my chatter. And when it was all done, copy editor Steven Boldt came along and made it even better with his careful eye and subtle editing.

Dick Anderegg, Hal Bonawitz, Stanley W. Cloud, Rocky Crandell, Bob Decker, Murray Gart, Grant Heiken, Rick Hoblitt, David Nolan, John Power, Miriam Rabkin, James Stacy, Don Swanson, Barry Voight, and many others gave this book a careful read and then unleashed their honest comments. They made this a more readable and more accurate history. Dan Miller labored through several versions, collected photographs and maps, and guided me around St. Helens, allowing me to see the eruption through his eyes.

Hoblitt, Miller, and many others stood for long hours over Xerox machines to copy field notes, journals, letters, and official documents. Dan Miller's mother, Virginia, deserves special thanks for clipping news stories that mentioned her son and his volcanoes. Thanks, too, to Barry Voight, who provided me with data specially prepared for this book.

Thanks to *Longview Daily News* newspaperman Andre Stepankowsky for helping me get a feel for the culture and politics of the southwest corner of Washington State. And thanks to his editors for opening the paper's library to me.

Lots of people who made important contributions to volcanology during this time do not get the attention here that their work deserves. To tell this story, I decided to keep the number of "characters" to as few as possible. Perhaps a better writer could have found a way to pay tribute to all at the length their contributions deserved. They include Ray Wilcox (and his family) for his detailing the Parícutin eruption in the late 1940s and the refinement of a method of fingerprinting volcanic ash from different eruptions. Dick Janda had an unusual genius as a scientist and a manager; he did extraordinary work at St. Helens following the May 18 eruption, and at other spots including Ruiz, Redoubt, and elsewhere; and he rescued VDAP at a critical period. His early, tragic death from cancer cost geology unknown discoveries and a much-admired colleague. Tom Casadevall did extraordinary early work on gas chemistry and highlighted the dangers of volcanic ash to jetliners. Norm Banks had the vision and persistence to create the program that eventually became VDAP. VDAP's responses to volcanic eruptions worldwide endures as a tribute to his vision and determination. These people and dozens of others who are not mentioned here gave me an enormous amount of time and treasured material, for which I shall forever be grateful.

I want also to thank the families and friends of Harry Glicken and Dave Johnston, and especially Dave's close friend Chris Carlson, who shared so many memories with me. I was reluctant to contact these people, but in the end it was always productive. I was especially grateful for

Mrs. Johnston's comment after one long interview. She said, "It's been like a visit with Dave." For me, too.

And thanks to Dave Morton at the University of Colorado's Hazards Center Library in Boulder. And Dave Wieprecht at CVO and many others did marvelous jobs of collecting photographs for the book.

To all these people, I want to say I hope I have faithfully captured the human story about what was one of the most exciting and tragic times in this science.

I especially want to thank Chris Ogden for years of unsolicited advice and exceptional friendship. My old colleague, longtime friend, and best man, Robert Buderi, has always set the standard for me. Thanks for putting the bar so high. I want to thank the D. G. Lunch Group, including Jeff Birnbaum, Ted Gup, Stanley Kayne, and Mike Riley for their friendship and support. Wendy "Gwendolyn" King and Brian Doyle provided consistent good cheer. Don Collins and James Coburn gave me their technical support, also with much good cheer. And I would like to thank *Time* magazine's Washington bureau chief, Michael Duffy, who may be the best boss any reporter ever had.

Most important, I want to thank my best friend, who is also my wife, Kristin. When we married, I knew she was a remarkable woman. But every year since, in so many wonderful and sorrowful corners of the world, she has become an even more lovely person than the one I loved the year before. I am a lucky dog. Thank you, K, for all my happiness.

Index

Note: *Italic page references indicate illustrations.*